"Awesome —— a must for QI specialists and a delight for non-specialists."

Paul de Groot
co-founder of dGB Earth Sciences

"[This] compilation contains many thoughts on foundation concepts, balanced with the caveats under which they apply. It reminds us of the old adage that 'all models are wrong, some are just less wrong than others', and that all measurements are uncertain. It admonishes us that we have a lot to learn and as such contributes to keeping rock physics a fresh and young discipline."

Douglas R Schmitt
Canada Research Chair in Rock Physics, University of Alberta

"Hall and Bianco have generated a readable book on rock physics by attracting some of the top names in our industry to contribute to this useful work."

Mike Bahorich
President of SEG, 2002–03

52 THINGS
YOU SHOULD
KNOW ABOUT
ROCK PHYSICS

EDITED BY MATT HALL & EVAN BIANCO

AgileLibre

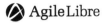

First published in 2016 by Agile Libre
Nova Scotia, Canada. *www.agilelibre.com*

Technical editors Matt Hall & Evan Bianco • *Managing editor* Kara Turner
Indexer Isabel Steurer • *Layout design* Neil Meister, MeisterWorks

Cover design electr0nika and Matt Hall.
The cover shows the fractured surface of Ketton stone, a Jurassic oolite from England. Drawing by Robert Hooke, as published in Schem. IX, Fig. 1 of *Micrographia* (1665).

We have done our best to ensure that the non-subjective parts of this book are factually accurate. If you find a typo or a factual inaccuracy please let us know at *hello@agilelibre.com*. While every precaution has been taken in the preparation in this book, the publisher, editors, and contributors assume no responsibility for damages resulting from the use of the information contained herein.

Library and Archives Canada Cataloguing in Publication

52 things you should know about rock physics / edited by
Matt Hall and Evan Bianco.

Includes biographical references and index.
ISBN 978-0-9879594-5-4 (paperback)

1. Rocks. 2. Petrology. 3. Rock mechanics. 4. Geophysics.
I. Hall, Matt, 1971-, editor II. Bianco, Evan, 1982-, editor
III. Title: Fifty-two things you should know about rock physics.

QE431.5.F54 2016   552   C2016-902585-3

Who we are

Agile Libre is a small independent publisher of technical books in Nova Scotia, Canada. Our passion is for sharing, so our books are openly licensed and inexpensive to buy. We hope they inspire you.

Where to get this book

This book is part of a series; the other books are *52 Things You Should Know About Geophysics*, *52 Things You Should Know About Geology* and *52 Things You Should Know About Palaeontology*. You will find these book for sale at *agilelibre.com* and also at Amazon's various stores worldwide. Professors, chief geoscientists, managers, gift-givers: if you would like to buy more than 10 copies, please contact us for a discount at *hello@agilelibre.com*.

About open licenses

Colophon

This book was laid out on a Mac using Adobe InDesign. The cover typeface is Avant Garde Gothic and the text typefaces are Minion and Myriad. The figures were prepared in GIMP and Inkscape. It is printed and distributed through Amazon's CreateSpace publishing platform.

Contents

Alphabetical

Contents

By theme

EXPLORATION
EXPLORATION • INTEGRATION

WORKFLOW
WORKFLOW • INTERPRETATION

UNCERTAINTY
UNCERTAINTY • PITFALLS

MODELLING
MODELLING • THEORY • QUANTITATIVE

CONSTITUENTS
SHALE • FLUID

BEHAVIOUR
ANISOTROPY • MODULI • GEOMECHANICS

Introduction

Technical books often start with a definition, usually something from a famous dictionary. It's unlikely any famous dictionary has an entry for 'rock physics', but even if it did I doubt you'd find anyone in the field who'd agree with it.

So let's excuse ourselves from a definition, and instead embrace the diversity. We've collected essays from 46 authors that get into every corner of the discipline, leaving, as they say, no rock unturned. There are essays full of mathematics, and others full of history; some with advice, and others with warnings. If your favourite subject was not covered, please let us know so we can invite you to contribute to the next volume!

The eclectic mixture is a function of the breadth of experience and interests of the writers. As well as pieces by no fewer than four past presidents of the Society of Exploration Geophysicists — and likely several future ones — you will find essays by industry, academic, and government scientists at all stages of their careers. Each of them has shared something important to them and, we hope, to you. Many of them are so unexpected and unintuitive, they have forever changed how we (the editors) look at log and seismic data.

Notwithstanding all this, we must recognize that diversity remains a problem for us. This book, like our entire profession, suffers from a discouraging preponderance of white male voices. We continue to work on this but it's clear that we don't know how to solve it. If you do, please get in touch.

Before you get into the essays, there are some things you need to know about the *Things You Should Know* series, in which this is the fourth book (see the back of the book for information about the others). All of the written words and, unless otherwise labelled, the figures too, are open content. What does that mean? In short, it means anyone is free to share and reproduce the content in this book, for any purpose and without permission — provided you give attribution to the author or authors. Most books are not like this! By the way, if you do re-use any of the work, I am certain the author would appreciate hearing about it.

If you enjoy this book, please share it with others. If you want lots of people to read it — your bosses or your team or your colleagues or your students — get in touch. We can make it more affordable. Times are hard in applied geoscience, especially in petroleum-related fields. The downturn had a profound effect on the schedule of this book, as our energy was diverted to other more pressing matters. But we don't want the lull to get in the way of professional geoscientists everywhere enjoying and learning from its contents.

We're thrilled to see this book get out into the world. We hope it inspires you to learn and keep learning. Hunt down the references in the essays you enjoy. Seek out the wonderful books from the centre of excellence at Stanford — *The Rock Physics Handbook* (Mavko et al., 2009, 2nd ed., Cambridge) and *Quantitative Seismic Interpretation* (Avseth et al., 2010, Cambridge) deserve a place on every practitioner's shelf. And rock physics lends itself so well to computation and data analysis — see Alessandro Amato's excellent tutorial on seismic petrophysics with Python in *The Leading Edge*. Together with the MATLAB code from the Mavko and Avseth books (op. cit.), there is plenty for the beginner and the experienced programmer alike.

The topic may be impossible to define, but we hope this book — this celebration of the field — will be your new favourite answer to the question, 'What is rock physics?'.

Matt Hall & Evan Bianco
Nova Scotia, August 2016

References

Amato, A (2015). Geophysical Tutorial: Seismic Petrophysics, Part 1. *The Leading Edge* **34** (4). DOI: 10.1190/tle34040440.1.

Amato, A (2015). Geophysical Tutorial: Seismic Petrophysics, Part 2. *The Leading Edge* **34** (6). DOI: 10.1190/tle34060700.1.

A primer on poroelasticity

Brian Russell

Many of the equations that we learn in our introductory geophysics classes make the assumption that the earth is perfectly elastic. This is a reasonable assumption to start with, but the reservoir rocks that we deal with in exploration and production geophysics are actually poroelastic, meaning that the propagation of seismic waves is influenced not only by the rock frame, but also by the pores in the rock and the fluids filling the pores. Poroelasticity theory was initially developed by Maurice Biot (1941) and then independently by Fritz Gassmann (1951), so we now refer to the Biot–Gassmann theory of poroelasticity. The original papers by Biot and Gassmann concerned static poroelasticity, or what happens when you 'squeeze' a porous fluid-filled rock. Biot's later work concerned dynamic poroelasticity, which concerns transient effects, such as seismic waves as they pass through the porous rock. By discussing the elastic constants themselves we can make several important observations about the effects of static poroelasticity on seismic velocities in the reservoir.

By applying stress (pressure) to a rock sample and measuring the resulting strain (deformation), rock physicists define five basic elastic constants:

- **Bulk modulus**, K, the ratio of the sum of principal stresses to the sum of principal strains.
- **Young's modulus**, E, the ratio of the first principal stress to the sum of principal strains.
- **Shear modulus**, μ, the ratio of x-z shear stress to twice the x-z shear strain.
- **Poisson's ratio**, ν, the ratio of the third principal strain to the first principal strain.
- **The first Lamé constant**, λ, the sum of the bulk modulus plus two-thirds of the shear modulus.

Note that μ is also the second Lamé constant and that the two Lamé constants can be derived from the propagation of P- and S-waves in an elastic and isotropic earth. The S-wave velocity is the square root of the ratio of the shear modulus divided by the density, or

$$V_s = \sqrt{\frac{\mu}{\rho}}$$

We know the density of a porous gas-filled sandstone reservoir is lower than in a wet reservoir. Biot–Gassmann theory tells us that the shear modulus is unaffected by the fluid filling the pores. This leads us to the interesting observation that the S-wave velocity is actually higher in a partially or fully gas-filled reservoir than it is in a wet reservoir.

The P-wave velocity is dependent on two elastic constants and density, and can be written in two separate ways as

$$V_{\mathrm{P}} = \sqrt{\frac{K + \sfrac{4}{3}\mu}{\rho}} = \sqrt{\frac{\lambda + 2\mu}{\rho}}$$

The fact that the P-wave velocity can be defined two different ways reflects the fact that K, μ and λ are related to each other, as discussed above. Unlike the shear modulus, the bulk modulus and first Lamé constant are dependent on the fluid filling the pores, and the Biot–Gassmann relationship can be written very succinctly in Biot's notation for these moduli as:

$$K_{\mathrm{sat}} = K_{\mathrm{dry}} + \alpha^2 M \quad \text{and} \quad \lambda_{\mathrm{sat}} = \lambda_{\mathrm{dry}} + \alpha^2 M,$$

where the $_{\mathrm{sat}}$ subscript refers to the saturated rock, $_{\mathrm{dry}}$ refers to the dry rock (with the fluids drained out), and α and M are related to the fluid in the pores. Thus, by substituting these two equations into the expressions for V_{P} and V_{S}, we can separate out the dry component of each modulus from a second term that is identical in both cases and is related to the fluid content of the rock. This has been applied to both inverted seismic results and AVO reflectivity terms (Russell et al. 2003, Russell et al. 2011). Recalling that impedance is density times velocit, for the inverted P impedance, I_{P}, and S impedance, I_{S}, this leads to the relationship

$$\alpha^2 M = I_{\mathrm{P}}^2 - (V_{\mathrm{P}} / V_{\mathrm{S}})_{\mathrm{dry}}^2 I_{\mathrm{S}}^2,$$

Furthermore, when $(V_{\mathrm{P}}/V_{\mathrm{S}})_{\mathrm{dry}}^2$ equals 2 it leads to the lambda-mu-rho (LMR) theory of Goodway et al. (1997), which is a subset of this more general theory.

References

Biot, M (1941). General Theory of Three-Dimensional Consolidation. *Journal of Applied Physics* **12**, 155–164. DOI: 10.1063/1.1712886.

Gassmann, F (1951). Über die Elastizität poröser Medien. *Viertel. Naturforsch. Ges. Zürich* **96**, 1–23. Reproduction by permission of Naturforschende Gesellschaft in Zürich, online at ageo.co/gassmann-de. English translation in *Classics of Elastic Wave Theory*, SEG Geophysics Reprint no. 24, online at ageo.co/gassmann-en.

Goodway, W, T Chen, and J Downton (1997). *Improved AVO fluid detection and lithology discrimination using Lamé petrophysical parameters*. Extended Abstracts, SEG, 67th Annual International Meeting, Denver.

Russell, B, K Hedlin, F Hilterman, and L Lines (2003). Fluid-property discrimination with AVO: A Biot–Gassmann perspective. *Geophysics* **68**, 29–39. DOI: 10.1190/1.1543192.

Russell, B, D Gray, and D Hampson (2011). Linearized AVO and poroelasticity. *Geophysics* **76**, C19–C29. DOI: 10.1190/1.3555082.

Acoustic emission

Michael King

The 50-mm side cubic specimen of fine-grained Fontainebleau sandstone (illustrated below), isotropic in its mechanical properties, was subjected in a polyaxial testing machine to stress conditions leading to failure. In order to achieve this, the minor principal stress was maintained constant at a low level (σ_x) while the two other principal stresses were increased in unison until failure occurred. In order to prevent the formation of shear stresses at the rock–steel platen interfaces in the testing machine, magnesium plates 5 mm thick were inserted between the testing machine platens and each of the six faces of the rock specimen. Magnesium had previously been shown to match the elastic properties (Young's modulus divided by Poisson's ratio) of both this particular rock and the steel platens. One of the magnesium plates is shown on the right in the illustration. As predicted theoretically for an isotropic elastic material, a number of failure planes were formed perpendicular to the minor principal stress rock and the steel platens:

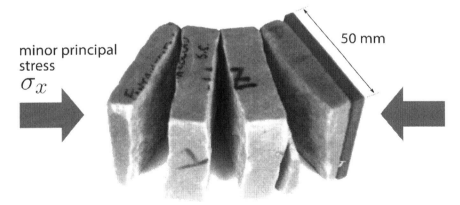

minor principal stress σ_x

50 mm

Consider now the case that if, at the start of the experiment, the minor principal stress had been set at a level a little greater than σ_x above, and the two other principal stresses increased to the values causing failure, as before. The rock specimen will remain intact. Now increase the pore fluid pressure ρ_f until failure takes place. This occurs when the effective stress ($\sigma_x - \rho_f$) reaches a value similar to that of the minor principal stress at failure considered above. The effective stress criterion is particularly important for those cases in which the pore fluid

…During the initial stages of fracture the [acoustic emission] events were predominantly tensile for all specimens…

is employed to fracture the rock. Failure occurs at the point the *effective stress* reaches the value of the minimum principal stress.

During the failure process for a number of porous sandstones — with porosities ranging from 4 to 25 percent — we have tested, acoustic emission (AE) events associated with the formation of the fractures were monitored by 24 small diameter AE sensors (frequency range to 2 MHz) located strategically in the platens of the polyaxial testing machine. Of these sensors, eight were employed also as compressional-wave transmitters to establish the velocity structure throughout the rock specimen at specific intervals during the test. This technique provides the means for identifying the position of the AE events. Moment tensor analysis of these results established the time at which each of the through-going fractures propagated and the character of the failures. This analysis indicated that during the initial stages of fracture the AE events were predominantly tensile for all specimens, whereas at later stages, approaching failure, they became predominantly shear in nature.

This type of analysis has applications in studies of fracturing of rock due to the injection of fluids in a number of practical field situations. These include hydraulically fracturing from boreholes intersecting a rock formation, such as low-permeability shale for producing oil or gas, or deep hot dry rocks to produce steam for energy production. An additional case applies to the undesirable production of microfractures caused by heating the rock adjacent to nuclear waste stored deep underground, which might permit groundwater to access the waste containers. In all these cases it is important to establish the extent to which the fractures spread by monitoring their propagation by AE studies. A number of studies have been reported in the scientific literature of all three types of application over the past few years.

References

Rubinstein, J, and A Mahani. (2015). Myths and Facts on Wastewater Injection, Hydraulic Fracturing, Enhanced Oil Recovery, and Induced Seismicity. *Seismological Research Letters*, **86** (4). DOI: 10.1785/0220150067.

Ask your data first, not your model

Paola Vera de Newton

Understanding rock types and their microstructure is almost an impossible task if we don't exhaust all available data. From basic logs to detailed core analyses, rock physicists build mathematical models from what we know about the in situ reservoir conditions. If the model doesn't reproduce the measured data, your data is telling you the model is wrong, your assumptions are wrong, or possibly both.

Quality control of log and core data is the most important step for a successful rock physics diagnostics and modelling study. Log editing is needed, in many cases, due to poor measurements in the borehole. This is another example of data revealing important pieces of information. Do we have reservoir damage? How much do we trust the logs? Is there a geological explanation for what I'm seeing? A very good piece of advice I received from my mentors at Rock Solid Images was to ask yourself what the data tells you first, and then question the model. The process of diagnosing reservoir and non-reservoir zones involves interpreting lithology, and applying some knowledge of microstructure and geological setting. Key inputs in any rock model include mineral constituents, saturation, pressure, temperature, and fluid data. Rocks are complex systems.

Here are some suggestions to help prepare for rock physics diagnostics and modelling:

- Evaluate the quality of your logs. From a simple depth shift to an invasion correction, this practice can help reduce uncertainty in the rock model calibration.

- Identify your reservoir and non-reservoir zones, and their properties. Yes, this means a full mineralogy analysis. If no core is available, dig into your mudlog and any well reports. They can take you a long way in cases where data is sparse.

- Start by running simple isotropic models. Conventional and elegant, Hashin–Shtrikman bounds provide a good sense of end members for bulk and shear moduli. From conventional sandstones and shaly reservoirs to complex carbonates, these bounds offer rock physicists a quick understanding of how possible constituents vary within the matrix.

Matching your data doesn't mean you have a perfect model,
because there is no such thing.

- Incorporate your geological understanding of the zone of interest into the modelling. This can be done by defining parameters related to various processes such as diagenesis and compaction.

- Model the in situ response. Matching your data doesn't mean you have a perfect model, because there is no such thing. The earth is too complex to be simplified using rock physics models. One would need an infinite number of models to replicate every layer in the reservoir. Instead, interpret your results even when your model fails by using the concept of elastic equivalency. This concept assumes we can only calculate bulk responses of sediment layers due to tool resolutions. Therefore we need to ensure our rock physics model replicates — with some degree of confidence — such bulk responses by properly understanding the in situ elastic properties and the petrophysical model.

- Perturb the rock with care: your model also has physical limitations. While modelling, use those rock property combinations that seem most likely. Start by isolating each effect, and then combine them.

- Don't forget to generate a rock physics template! This tool can summarize numerous models and give geoscientists a good theoretical understanding of rock property changes for a given reservoir.

Attributes on the rocks

Sven Treitel

When seismic attributes became popular in our industry, I was skeptical at first. I eventually realized that with proper use, they could be of great value to the seismic interpreter. On the other hand, our understanding of just how such seismic attributes relate to rock properties was — and continues to be — incomplete. That's okay; exploration seismology is not an exact science — we are comfortable using methods that work even when we do not fully understand why or how they work. Indeed, I first learned about attributes from my late friend Tury Taner, who was one of the earliest proponents of these ideas, and is perhaps no accident that Tury was both an accomplished artist as well as an outstanding scientist.

Because seismic attributes are determined entirely from seismic recordings, it stands to reason that many attributes should be sensitive to the physical properties of the rocks through which the seismic waves travelled — some more, some less, some not at all, depending on which attributes are used. How might we go about studying the problem?

I would envision an investigation to consist of two components, one theoretical, the other experimental.

Theoretical

A wealth of mathematical models exists relating rock properties such as anisotropy, fluid content, porosity, fracture characteristics, and density to synthetic traces. The traces will be functions of one or more of these parameters. A set of attributes could then be calculated from the synthetic data to identify those most responsive to a given rock property, and under which recording conditions this response might occur. In other words, such calculations could help establish which attributes are more sensitive to a particular rock property, and which are less so.

Empirical

While some might chose to call it a day and stop here, this may be premature: we write down equations which we think describe a physical characteristic we are trying to capture with an attribute, but of course there is no assurance

Exploration seismology is not an exact science — we are
comfortable using methods that work even when
we do not fully understand why or how they work.

that what we have written down fully simulates an actual physical process taking place below ground. A second component should then involve laboratory measurements on both natural and artificial rock samples. A serious problem here is that frequencies used in the laboratory are much higher than those used in field data acquisition, meaning that corrections for such scaling problems would have to be made. The point is that numerical models used for computer simulations should be based on observed laboratory data in addition to purely theoretical considerations.

In former times, a project of this kind would have been tackled by one of the major oil company labs. Today, the best place for it would probably be at a university, as a consortium funded by industry. But wherever it happens, I am convinced of the need for a more thorough understanding of the dependence of seismic attributes on the rocks.

Be careful when fitting models

Vedad Hadziavdic

There is a lot of geological and physical insight in making a rock physics model fit your data. However, forcing a model to fit by choosing parameters which do not reflect the underlying physics or geology, leads to bad decisions.

The models we use in rock physics are approximations. The mathematical complexity of modelling composite materials is so high that even if it were possible to replicate the internal geometry of rocks, it would result in algorithms which would be unpractical to apply in everyday routines. That's why we make simplifying assumptions about the nature of contacts between grains, the shape of pores, how rocks lose porosity with time, etc. Fortunately, despite these assumptions, the models are still useful. Up to a point.

Let me illustrate this by an example. The fact that our models over-predict shear velocities measured in the laboratory has been attributed to several potential issues, and fixed by introducing 'shear reduction factor' (SRF) in all standard software. Let's focus on one possible explanation in particular: the (wrong) assumption that grains will never slip regardless of the magnitude of the shear force in the Hertz–Mindlin model. By multiplying the shear modulus by a factor between 0 and 1, you introduce a 'slip' and make things right. The question is, what value of SRF can you live with? Let's say your target is at 1500 m depth. Does it make sense to apply 0.7 SRF? Probably, but what if the same sand is at 2500 m depth? Some values will require quite special physical assumptions. SRF = 0 means no shear resistance at all, which is not realistic. Small values can arise but probably not in sedimentary rocks. For example, SRF = 0.1 means the grain contacts slip at quite small shear stress. In some software packages, you will find SRF options in models for cemented sands too. How much slip would you expect in those cases?

A reasonable question to ask now is: I have tried everything and my model still does not fit, what do I do next? This brings us to the essence of fitting rock physics models. Not being able to fit a model usually means that we either do not understand, or are not able to model, the depositional environment, burial history, diagenetic alterations, or some physical conditions like pressure, temperature, or the internal geometry of rocks. Clay minerals in the frame will push your model in the same direction as SRF, some more than others. Based on input from the sedimentologist, you will know whether feldspar or smectite are more likely to be found in your rock, and they will affect your model in different ways. This goes for volume fractions too. If you need a large amount of feldspar in your frame in order to fit a model, it means you have an immature sand. What does your sedimentologist think of that?

In the end, if you have tried everything and an unrealistically low value of SRF in a cemented sand is the only thing that can rescue the model, remember that you have one more option — admitting that there is something you don't understand despite your best effort. Bringing that uncertainty into the decision process will not hurt you and your company, quite the opposite. It will strengthen the standing of rock physics as a useful input in the decision process. Your colleagues will simply trust you more next time you tell them that the model *does* fit the data.

Beware of shortcuts

James G Berryman

When I joined the oil industry workforce in the late 1970s, I had four years of undergraduate, six years of graduate, and one year of post-doctoral training in physics, math, and computing. But I had no training or practical experience with either geology or geophysics. This situation was not uncommon at the time because few universities had created the applied geophysics and geology programs that many schools now have.

On my first day at the job, I was given an office and presented with a large stack of photocopied technical papers to read and absorb. The main task I was assigned was to become the local expert on seismic wave propagation in media saturated with fluid — oil, gas, water or air. My colleagues were very helpful, but not necessarily experts on these issues — otherwise, why hire and train a new guy?

The papers I found most useful were often the ones written by Maurice Biot, but I also kept running across references to papers by Fritz Gassmann. The one quoted most often was 'Elastic waves through a packing of spheres' in *Geophysics* (1951), so when people talked to me about Gassmann and Gassmann's equation I assumed that this was the paper that contained the most important results. It took me a long time — and some confusion — to realize that the more important paper (while referenced in that same *Geophysics* journal) was actually published in an obscure place and in German. With some friends, I translated it! The translation is now available in *Classics of Elastic Wave Theory*, SEG Geophysics reprint no. 24, and online at *ageo.co/gassmann-en*.

The point is that I made the mistake at the time of assuming that the paper most often cited was the most important paper, when in fact it was the first reference in that paper that was really the more significant one.

So, beware of shortcuts, they can cost a lot of time and cause much confusion.

Vierteljahrsschrift der Naturforschenden Gesellschaft in Zürich

unter Mitwirkung von

A.U. DÄNIKER, P. FINSLER, H. FISCHER, A. FREY-WYSSLING, H. GUTERSOHN, P. KARRER, B. MILT
P. NIGGLI, P. SCHERRER, H. R. SCHINZ, FR. STÜSSI und M. WALDMEIER

herausgegeben von

HANS STEINER, ZÜRICH 7

Druck und Verlag: Gebr. Fretz AG., Zürich

Nachdruck auch auszugsweise nur mit Quellenangabe gestattet

Jahrgang 96	HEFT 1	31. März 1951

Abhandlungen

Über die Elastizität poröser Medien

Von

F. GASSMANN, Zürich

(mit 3 Abbildungen im Text)

I. Gegenstand der Untersuchung

(1) Es gibt Stoffe, die in einem mehr oder weniger weiten Spannungs-bereich als ideal elastisch betrachtet werden können, d. h. bei denen zwischen dem Spannungszustand und dem zugehörigen Verformungszustand mit grosser Genauigkeit Proportionalität besteht. Dazu gehören bei Zimmertemperatur z. B. die harten Metalle, die meisten Mineralien, die reibungslosen Flüssig-keiten.

(2) Es gibt andere Stoffe, deren Verhalten erheblich vom ideal elastischen abweicht, wie z. B. die plastischen Stoffe, bei denen ein Spannungszustand irreversible Verformungen bewirkt. Zu ihnen gehören auch mehrphasige Systeme, wie etwa feste poröse Körper oder lockere Aggregate von Körnern, wobei die Poren oder Zwischenräume mit Flüssigkeiten oder Gasen gefüllt sind. Bei der Beanspruchung solcher Systeme treten irreversible Zustands-änderungen auf, etwa weil die Porenflüssigkeit wandert oder die Körner eines Aggregates ihre Packung verändern.

(3) Nun besteht aber die Tatsache, dass auch solche nicht ideal elastische Systeme sich wie ideal elastische verhalten, wenn der Spannungszustand kleine Schwankungen um einen gegebenen mittleren Spannungszustand aus-führt; d. h. die Verformungsschwankungen, die durch die kleinen Spannungs-schwankungen bewirkt werden, sind mit grosser Genauigkeit reversibel und proportional zu ihnen. Solche Spannungsschwankungen treten z. B. auf, wenn

Page 1 of the original Gassmann paper in German.

References & acknowledgments

Gassmann, F (1951). Über die Elastizität poröser Medien. *Viertel. Naturforsch. Ges. Zürich* **96**, 1–23. Reproduction by permission of Naturforschende Gesellschaft in Zürich, online at ageo.co/gassmann-de. English translation in *Classics of Elastic Wave Theory*, SEG Geophysics Reprint no. 24, online at ageo.co/gassmann-en.

Gassmann, F (1951). Elastic waves through a packing of spheres. *Geophysics* **16** (4), 673–685. DOI:10.1190/1.1437718.

Build a good low-frequency model

Per Avseth

Seismic inversion for absolute values of elastic parameters (e.g. acoustic impedance and V_p/V_s) requires low frequency information from a model or other source of data that contains sufficiently low frequencies. It's the low frequencies that scale our inversion data to the right ballpark. Moreover, the low-frequency model serves as a starting model that is iteratively updated until the error between the synthetic data and the real data is minimized. Also, the low-frequency model contains information about geological structures and compaction trends, and therefore constitutes an important constraint on the quantitative seismic interpretation (e.g. Ray and Chopra 2016).

Normally, the low-frequency model is obtained from smoothing sonic and density well-log data. In between wells, it is common to interpolate the smoothed values. If wells are scarce, interval velocities derived from stacking velocities can also be used to build low-frequency models for seismic inversion. But seismic velocities are famously unreliable, so a third alternative is to use rock physics depth trends. These can be based on empirical compaction trends, or from combined basin modelling and rock physics modelling. The latter is a focus of recent research (e.g. Brevik et al. 2014; Dræge et al. 2014; Avseth et al. 2016), and can be very useful in areas with poor well control where one needs to extrapolate away from existing wells.

There are other benefits to rock physics models. They can extrapolate low-frequency models to capture geology not encountered by existing wells — for example in an area with both synclines and anticlines, where wells are predominantly situated on the anticlines. It is important that the low-frequency trend used in the inversion is in agreement with the trend in the training data used for the subsequent facies classification. Also bear in mind that the low-frequency information that is expressed in weak contrast reflectivity approximations (like Aki and Richards) is defined by the average of two layers; hence the low-frequency trend used in seismic inversion should never be a pure shale trend! It should be the average of the background material and the 'unknown' rock in the second layer. A good approach is to make the low-frequency trend the average of water-saturated sand and water-saturated shale. However, sands and shales can be interbedded at a scale not resolved by the seismic, and then strictly speaking the low-frequency model should honor the correct net-to-gross, which is normally unknown away from well-control. Alternatively, one way to avoid this 'average rock' low-frequency trend is to introduce low-frequency trends or depth trends for separate lithologies; these can be used to invert for lithologies first, followed by elastic properties constrained by the given facies. This approach was described by Rimstad et al. (2012) and Kemper and Gunning (2014).

A good way to assess the quality of the low-frequency model is to crossplot the model in a rock physics template, and then perform classification of the low-frequency model using the same training data as used in the classification of the final inversion data. This may be an iterative approach, where the prior model needs to be updated after the inversion is conducted, to obtain a refined version that complies better with the selected rock physics template.

What if well-log data used to build a low-frequency model includes a hydrocarbon zone? It may be advisable to fluid substitute the hydrocarbon zone to water before smoothing the well-log data. Then we avoid adding information from the model about the presence of hydrocarbons, and we can check if the seismic inversion will predict the hydrocarbon zone, even though the low-frequency model assumes water-saturated rocks only. It is also recommended to exclude some wells, both in the estimation of wavelets and the low-frequency model and perform rigorous blind tests during the inversion.

The bottom line is that it is important to ensure good integration between geologists and geophysicists when preparing the low-frequency model for an absolute seismic inversion, and rock physics serves as a good platform for this integration.

References

Avseth, P, and I Lehocki (2016). Combining burial history and rock-physics modeling to constrain AVO analysis during exploration. *The Leading Edge* **35** (6), 528–534. DOI: 10.1190/tle35060528.1.

Brevik, I, T Szydlik, MP Corver, G De Prisco, C Stadtler, and HK Helgesen (2014). Geophysical basin modeling: Generation of high quality velocity and density cubes for seismic imaging and gravity field monitoring in complex geology settings. *SEG Technical Program Expanded Abstracts* **2014**, 4733–4737. DOI: 10.1190/segam2014-0444.1.

Dræge, A, K Duffaut, T Wiik, and K Hokstad (2014). Linking rock physics and basin history — Filling gaps between wells in frontier basins. *The Leading Edge* **33** (3). 240–242, 244–246. DOI: 10.1190/tle33030240.1.

Kemper, M, and J Gunning (2014). Joint impedance and facies inversion — Seismic inversion redefined. *First Break* **32**, 89–95.

Ray, AK, and S Chopra (2016). Building more robust low-frequency models for seismic impedance inversion. *First Break* **34**, 47–52.

Rimstad, K, P Avseth, and H Omre (2012). Hierarchical Bayesian lithology/fluid prediction: A North Sea case study. *Geophysics* **77** (2), B69–B85. DOI: 10.1190/geo2011-0202.1.

Consider the error in the measurements

Kyle T Spikes

When we use a rock physics model to help interpret a well-log dataset, we need to remember that those data contain errors. We recognize three sources of error in general for such analysis:

1. Error in the model.
2. Error in the data-to-model match.
3. Error in the data.

Error in the model can be understood based on its underlying assumptions and inputs into the model. Model-to-data misfits can be calculated using some type of norm or misfit. However, error in the data should be addressed with the experiment in mind, which is the focus of this essay.

Consider compressional and shear sonic log data. The experimental error in the measurements comes from many sources including the effects of the tool in the borehole, the drilling mud, the invaded zone, and borehole rugosity. Velocities, or slownesses, are picked using a semblance plot of frequency versus travel time computed from the recorded waveforms. These velocities correspond to seismic refractions along the borehole wall. We assume that the refraction propagates through the virgin formation. Furthermore, we assume that this velocity corresponds to a single homogeneous layer, but the length of the tool can be much larger than heterogeneities and thin laminations. The velocity at the lowest frequency and highest semblance value is most often automatically picked for both P- and S-waves. However, these picked velocities can be non-unique, particularly for the dispersive S-wave. Therefore, additional error can result from picking the incorrect value on the semblance plot.

With these errors in mind, consider North Sea siliciclastics that are assumed isotropic. We typically can discriminate the clean, high porosity hydrocarbon sands from the shaley sands and the shales by looking for trends in a domain such as in the figure. Only the clean and shaley sands are shown. The stiff sand model is shown in gray scale, where light gray is high clay content. The black lines are for constant values of modelled porosity, with values from 20 to 35 percent indicated. The oval indicates the cleanest, highest porosity sands with hydrocarbons. Lower porosity hydrocarbon sands with higher clay content have higher V_P/V_S. Open diamonds correspond to brine-saturated sands and closed diamonds to hydrocarbon sands, fluid substituted to 100 percent brine. The brine-saturated sands tend to have higher impedances. The clean, high porosity sands have the smallest V_P/V_S and acoustic impedance (I_P) between 5.00 and 5.25 km/s g/cm³, marked by the oval.

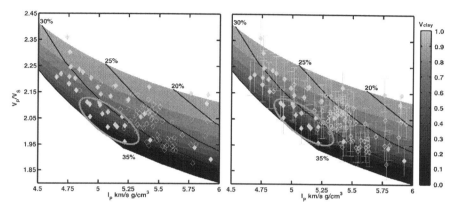

The domain is V_P/V_S as a function of I_P. Assume that the error in the model is negligible.

Next, let us add error bars to the V_P/V_S data as in the right-hand side of the figure — which shows the same model and the same data. Error bars for I_P also exist, but they are less significant than the error in V_P/V_S. Calculations of the total error in V_P/V_S result in about ±3 percent error based on 1 and 2 percent error in V_P and V_S, respectively. Immediately noticeable is that error bars obscure the trends we saw on the left. Most of the data points associated with the clean, high porosity sands easily could be interpreted as lower porosity and/or more shaley sands.

The error in the data should be considered when identifying the trends or relationships in elastic properties associated with reservoir sands. This error is important to remember during the analysis of well-log data and when applying such relationships to field seismic data. Ultimately, we must quantify all three sources of error to assess properly the goodness of fit and the accuracy of the interpretation.

De-risking low-saturation gas, part 1

K M (Chris) Wojcik

Low-saturation gas is commonly referred as 'fizz water' or 'fizz gas'. These terms are rather poorly defined and misleading (Han and Batzle, 2002). Zhou et al. (2005) proposed that 'fizz' refers to a gas reservoir with gas saturation less than 25 percent, while Han and Batzle (2005) suggested that a correct term for fizz gas is 'residual gas'. A common characteristic of fizz of any flavour is that the rock properties may be very similar to reservoirs with commercial hydrocarbons, so they are difficult to de-risk geophysically.

However, after more than two decades of research, the community has failed to establish methods to discriminate between commercial hydrocarbons and low-saturation gas: there is no high-confidence approach based on standard P-reflectivity seismic data (e.g. O'Brien 2004; Singh 2014). The solution to de-risking low-saturation gas tests even the most advanced geophysical methods and is based on a thoughtful, scenario-based integration with geology at basin, play, and prospect scales.

Systematics of low-saturation gas

There is a remarkable similarity in the occurrences of low-saturation gas in Tertiary deepwater turbidites. Usually, low-saturation gas is found in relatively shallow intervals, but the examples behind this summary include low-saturation gas reservoirs as deep as 3250 m below the seafloor with temperatures up to 80°C and pressures up to 62 MPa (cf. Simm and Bacon 2014). Generally, the occurrences of gas with non-commercial saturation can be classified into two categories as follows.

1. **Low-saturation gas (LSG).** Low-saturation gas *sensu stricto* is any amount of gas in very low net-to-gross, laminated sandstone reservoirs and in silty mudstones. The presence of LSG in the system does not imply the presence of a trapping geometry. Gas saturation in the LSG case may reach 40 to 50 percent with net-to-gross of the host rock ranging from 20 to 30 percent. LSG is commonly characterized by low- to high-amplitude soft events with class 3 AVO, but these anomalies show no structural conformance. LSG-related amplitudes are generally splotchy and may occur on- and off-structure. An accurate description of the related seismic response is 'stratigraphic highlighting'.

After more than two decades of research, the community has failed to establish methods to discriminate between commercial hydrocarbons and low-saturation gas.

2. **Residual gas.** True residual gas is found in well-developed reservoir sands. It is the immobile gas phase with saturations ranging from 15 to 20 percent remaining in the reservoir after a fully charged trap loses its integrity and leaks (blown trap). The leakage may by partial or full and trap integrity failure may be caused by cross-fault leakage or top-seal failure. Residual gas may bring about convincing amplitude and class 3 AVO anomalies, good amplitude switch-offs, flat spots, and other fluid contact evidences.

Considering significant geological and geophysical differences between these two instances, it might be reasonable to shelve the fizz moniker and describe gas-related failures as either low-saturation gas (LSG), as described above, or, in a special case of trap integrity failure, residual gas.

Continued in part 2.

References

Han, D and M Batzle (2002). Fizz water and low gas-saturated reservoirs. *The Leading Edge* **21**, 1066–1074. DOI: 10.1190/1.1471605.

Han, D and M Batzle (2005). Diagnosis of "fizz-gas" and gas reservoirs in deep-water environment. SEG Annual Meeting, Houston, Expanded Abstracts, 1327–1331.

O'Brien, J (2004). Seismic amplitudes from low gas saturation sands. *The Leading Edge* **33**, 1236–1243. DOI: 10.1190/leedff.23.1236_1.

Simm R and M Bacon (2014). *Seismic Amplitude: An Interpreter's Handbook.* Cambridge University Press.

Singh, R J (2014). Exploration application of seismic amplitude analysis in the Krishna-Godavari Basin, east coast of India. *Interpretation* **2**, SP1-SP16. DOI: 10.1190/INT-2013-0197.1.

Zhou, Z, F J Hilterman, H Ren, and M Kumar (2005). Water-saturation estimation from seismic and rock-property trends. SEG Annual Meeting, Houston, Expanded Abstracts.

De-risking low-saturation gas, part 2

K M (Chris) Wojcik

Standard rock physics diagnostics for suspected low-saturation gas is a comparison of petrophysical data with expected properties of wet sandstone reservoirs and bounding shales using acoustic impedance versus compressional/shear velocity ratio (V_P/V_S) crossplots (cf. Avseth et al. 2005). The crossplots usually show significant contrast between LSG or residual gas occurrences and the observed or model-predicted properties of hydrocarbon-saturated sandstones. Low-saturation gas may be difficult to detect in other data (resistivity, neutron/density, or chromatograph readings while drilling) but the rock physics is usually robust.

Integration of geophysics and geology

Considering the geological systematics of low-saturation gas and the overall geophysical affinity between success and failure cases (commercial gas or oil versus LSG or residual gas), we need to approach de-risking of the potential low-saturation gas failures in an integrated context of charge setting, reservoir/seal stratigraphy, structural setting, trap geometry, and pressure regime.

Generally, shallow prospects (e.g. 1–3 km below the seafloor) in a setting with abundant evidence for gas presence (e.g. bottom-simulating reflectors, small hydrocarbon flags) will have higher low-saturation gas failure risk. Stratigraphic amplitude highlighting, paucity of fluid contact evidence, amplitude anomalies both in on-structure and off-structure position all point to a high risk of low-saturation gas. The most likely outcome is a mudstone/siltstone or a low net-to-gross laminated sand interval with non-commercial gas saturation. In practice, however, mudstone/siltstone and sand-rich facies can be found in the same intervals and co-occurrence of low-saturation gas, residual gas, and commercial gas is not that uncommon. Detailed stratigraphic assessment and accurate facies prediction is hence critical to correctly analyze amplitude patterns and arrive at realistic chance factors.

No guarantees

Presence of amplitude or AVO anomalies in only on-structure positions with robust fluid contact evidence (e.g. flat spots) and geological evidence for well-developed sands, does not guarantee success. The key risk in such a case is

*[It is] possible to correctly estimate the risk of low-saturation
or residual gas failure and, in many cases,
correctly predict the well outcome.*

residual gas caused by seal integrity failure. Accurate assessment of chance factors requires detailed trap analysis, including cross-fault juxtaposition and fault sealing capacity, and pressure analysis for cases with large column height. Complicated fault traps with many potential leak points, sand-on-sand juxtaposition, and partial fill inferred from amplitude distribution point to a high residual gas risk. Prospects with a large column height inferred from fluid contact and trap analysis may have high seal failure risk if the pressure gradient intersects the fracture gradient. Pressure-related failure in active settings, such as salt mini-basins or structures cored by mobile shales may result in a tilting of the palaeo gas–water contact pointing again at a high residual gas risk.

In spite of the similarity of seismic responses between low-saturation gas and commercial hydrocarbons, I believe that it is nonetheless possible to correctly estimate the risk of low-saturation or residual gas failure and, in many cases, correctly predict the well outcome. Geophysical evidence can be ambiguous, even with high-quality 3D seismic data, but a thoughtful, scenario-based integration with geology may significantly increase the confidence of our prediction and reduce the risk of low-saturation gas failure.

References & further reading

Avseth, P, T Mukerji, and G Mavko (2005). *Quantitative Seismic Interpretation*. Cambridge University Press.

A full version of this essay, with examples, details, and full analysis was published in the SEG/AAPG *Interpretation* journal: Wojcik, K M, E Gonzalez, and R Vines (2016). Derisking low-saturation gas in Tertiary turbidite reservoirs. *Interpretation* **4** (3), p SN31–SN43. DOI: 10.1190/INT-2016-0034.1.

Dry rock is not dry

Fuyong Yan

Gassmann fluid substitution is one of the most important tools for modelling the seismic responses of the reservoir rocks with different pore fluids. The Gassmann equations are often written in the form of:

$$\frac{K_{sat}}{K_m - K_{sat}} = \frac{K_{dry}}{K_m - K_{dry}} + \frac{K_f}{\varphi\,(K_m - K_f)} \quad \text{and} \quad \mu_{sat} = \mu_{dry}$$

where K_{sat} is the bulk modulus of the rock filled with water, K_m is the bulk modulus of the frame material, K_{dry} is the bulk modulus of the dry rock, K_f is the bulk modulus of the pore fluid, and φ is the porosity. μ_{sat} and μ_{dry} are the shear modulus of the fully saturated and the dry rock respectively.

The theoretical validity of Gassmann's equations is well founded. They are consistent with Biot theory, Kuster–Toksöz theory, and Hashin–Shtrikman bounds. But for reservoir rocks, we sometimes find that Gassmann's equations do not work very well. The mismatches are often blamed on ultrasonic frequency, but this is only partly responsible. To understand Gassmann theory and apply it properly, it is essential to have a good understanding of the 'dry' rock property.

The problem is that several explicit assumptions are made in the derivation of the equations. I would like to emphasize one assumption:

'The pores should be filled with a frictionless liquid or a frictionless gas.'
(Paragraph 31, Gassmann 1951).

A frictionless fluid is an ideal fluid that has zero viscosity, so the pore fluids in different pore shapes, such as a stiff pore and neighbouring cracks, can reach pressure equilibrium instantaneously. For reservoir rocks filled with frictionless fluid, there is no dispersion or attenuation caused by squirt flow; in contrast, the dispersion caused by Biot flow will occur at extremely low frequency. In practice, reservoir fluids are not frictionless and the pore fluids in the cracks of reservoir rocks do not satisfy the assumption of Gassmann's equations. Therefore this component of the pore fluids should be included with the frame material. For consolidated reservoir rocks, the effect of Biot flow is usually not significant and occurs at around ultrasonic frequency. So we can still use Gassmann's equations, but we should only consider the saturation effect of the pore fluids in the macropore system.

To understand Gassmann theory and apply it properly,
it is essential to have a good understanding
of the 'dry' rock property.

In his paper, Gassmann (1951) discussed the effect of hygroscopic water, also known as irreducible or bound water. He pointed out that elasticity of the frame may change through accumulation of hygroscopic water on the pore walls. Hygroscopic water distinguishes itself from normal pore fluid by its high density, high pressure, and smaller mobility. The attraction between the mineral surface and the hygroscopic water can be 1 GPa, whereas the pressure disturbance due to passing of seismic wave is in order of tens to hundreds of Pascals. Thus the hygroscopic water is immobile and needs to be regarded as part of the frame. For fine-grained rocks or rocks with high clay content, the volume of the hygroscopic water can be significant: 1 gram of clay mineral can have a pore-wall surface area of hundreds of square metres!

Gassmann (1951) used a numerical example to illustrate a sandstone with a total porosity of 17.1% when immersed in water, but some of the porosity is inaccessible to the water, so the apparent porosity was 13.3%. He suggested that the apparent porosity should be used and the inaccessible pores should be 'accounted for within the solid material' (Gassmann 1951). Pores should not necessarily be treated equally and as a whole entity.

In summary, we should not treat the 'dry' rock properties of reservoirs absolutely. Even lunar dry rock contains minute amounts of water (Robinson and Taylor 2014). Selection of the 'dry' rock properties is closely related to the effective porosity used in Gassmann's equations. The 'dry' rock should include pore fluids that are strongly attached to the pore walls and that have difficulty in reaching pressure equilibrium with the macropore fluid system. Gassmann's equations can only predict the saturation effect caused by the pore fluid in the macropore system.

References

Gassmann, F (1951). Über die Elastizität poröser Medien. *Viertel. Naturforsch. Ges. Zürich* **96**, 1–23. Reproduction by permission of Naturforschende Gesellschaft in Zürich, online at ageo.co/gassmann-de. English translation in *Classics of Elastic Wave Theory*, SEG Geophysics Reprint no. 24, online at ageo.co/gassmann-en.

Robinson, K L and G J Taylor (2014). Heterogeneous distribution of water in the Moon, *Nature Geoscience* **7**, 401–408. DOI: 10.1038/ngeo2173.

Elastic symmetry: isotropy

Evan Bianco

The concept of stress, a force acting per unit area, is essential in expressing the mechanical interaction between one part of a material body with another. In order to calculate the stress at a single point within a three dimensional material, we can consider the forces acting on three orthogonal planes intersecting at a point. Each plane can be subject to three components of stress, so in general, it takes nine components, a so-called stress tensor, to describe state of stress in an elastic solid:

$$\sigma = \begin{bmatrix} \sigma_{11} & \sigma_{12} & \sigma_{13} \\ \sigma_{21} & \sigma_{22} & \sigma_{23} \\ \sigma_{31} & \sigma_{32} & \sigma_{33} \end{bmatrix}$$

where σ_{ii} represent normal stresses acting perpendicular to each plane, and σ_{ij} represent the shear stresses acting parallel to each plane. This matrix is symmetric about the diagonal, which means it really only takes only six elements to describe the state of stress at a point, not all nine. As such, we can write the stress in shorthand, sometimes called Voigt notation: $[\sigma_{11} \; \sigma_{22} \; \sigma_{33} \; \sigma_{23} \; \sigma_{13} \; \sigma_{12}]^{\mathsf{T}}$.

As long as the amount of stress does not exceed the elastic limit, and cause permanent deformation, most materials obey Hooke's law: the measured strain is proportional to the amount of applied stress. In terms of the equation above, each of the six components of stress σ_{ij} is a linear combination of six components of strain ε_{ij}, scaled according to the matrix elements C_{ij}:

$$\begin{bmatrix} \sigma_{11} \\ \sigma_{22} \\ \sigma_{33} \\ \sigma_{23} \\ \sigma_{13} \\ \sigma_{12} \end{bmatrix} = \begin{bmatrix} C_{11} & C_{12} & C_{13} & C_{14} & C_{15} & C_{16} \\ C_{21} & C_{22} & C_{23} & C_{24} & C_{25} & C_{26} \\ C_{31} & C_{32} & C_{33} & C_{34} & C_{35} & C_{36} \\ C_{41} & C_{42} & C_{43} & C_{44} & C_{45} & C_{46} \\ C_{51} & C_{52} & C_{53} & C_{54} & C_{55} & C_{56} \\ C_{61} & C_{62} & C_{63} & C_{64} & C_{65} & C_{66} \end{bmatrix} \begin{bmatrix} \varepsilon_{11} \\ \varepsilon_{22} \\ \varepsilon_{33} \\ 2\varepsilon_{23} \\ 2\varepsilon_{13} \\ 2\varepsilon_{12} \end{bmatrix}$$

which are called the elastic stiffnesses. Although this equation looks rather gnarly, it just says 'stress is equal to something times strain'. The *something* — the stiffness matrix — describes how the material changes shape under different kinds of stress. To find all the elements in this 6×6 matrix is to find a complete elastic description of the rock. It can distinguish one kind of lithology or pore-fluid from another, and in combination with bulk density, tells us about the magnitude and variability of seismic wave speeds in different directions.

Fortunately, the stiffness matrix is also symmetrical along the diagonal which reduces the total number of elastic constants from 36 down to 21 at the most, depending on the degree of symmetry that the material has. In the simplest case, a material is isotropic, meaning it looks the same in all directions, and the stiffness matrix reduces to just two independent elastic constants: C_{11} and C_{12}:

$$\begin{bmatrix} C_{11} & C_{12} & C_{12} & 0 & 0 & 0 \\ C_{12} & C_{11} & C_{12} & 0 & 0 & 0 \\ C_{12} & C_{12} & C_{11} & 0 & 0 & 0 \\ 0 & 0 & 0 & (C_{11} - C_{12})/2 & 0 & 0 \\ 0 & 0 & 0 & 0 & (C_{11} - C_{12})/2 & 0 \\ 0 & 0 & 0 & 0 & 0 & (C_{11} - C_{12})/2 \end{bmatrix}$$

Casting this matrix in terms of the Lamé parameters λ and μ we get the well-recognized terms that show up in the equations for wave propagation and seismic reservoir characterization work:

$$\begin{bmatrix} \sigma_{11} \\ \sigma_{22} \\ \sigma_{33} \\ \sigma_{23} \\ \sigma_{13} \\ \sigma_{12} \end{bmatrix} = \begin{bmatrix} \lambda + 2\mu & \lambda & \lambda & 0 & 0 & 0 \\ \lambda & \lambda + 2\mu & \lambda & 0 & 0 & 0 \\ \lambda & \lambda & \lambda + 2\mu & 0 & 0 & 0 \\ 0 & 0 & 0 & \mu & 0 & 0 \\ 0 & 0 & 0 & 0 & \mu & 0 \\ 0 & 0 & 0 & 0 & 0 & \mu \end{bmatrix} \begin{bmatrix} \varepsilon_{11} \\ \varepsilon_{22} \\ \varepsilon_{33} \\ 2\varepsilon_{23} \\ 2\varepsilon_{13} \\ 2\varepsilon_{12} \end{bmatrix}$$

Specifically, the so-called P-wave modulus $\lambda + 2\mu$ shows up in the first three elements along the diagonal, and the shear modulus μ in the last shows up in the last three (see also *A primer on poroelasticity*, page 14).

The case of perfect elastic symmetry thus describes a system with two degrees of freedom. This is why AVO inversions take some flavour of near-offset seismic as one input and far-offset seismic as a second input and estimate two rock properties at each point in space. Rock physics methods may present seemingly obscure couplings of isotropic properties, such as V_P/V_S versus Poisson's ratio, or $\lambda\rho$ versus $\mu\rho$, but because of the high degree of symmetry for isotropic materials, it is trivial to cast any properties that you have into the two that are most meaningful to you. If you've got two, you've got them all — any two will do.

References

Bhatia, A and R Singh (1986). *Mechanics of Deformable Media*, IOP Publishing.

Elastic symmetry: anisotropy

Evan Bianco

A material is elastically anisotropic if it takes more than two elastic constants to describe how it will change shape to accommodate stress. It turns out that all minerals are anisotropic, and most rocks are as well. Truly isotropic media are quite rare, although sometimes geophysicists get away with the assumption.

The vast majority of anisotropy seen at the scale of seismic experiments deal with transversely isotropic (TI) models with the axis of symmetry in different orientations. Building on the mathematical description of stiffness tensors in *Elastic symmetry: isotropy*, we can describe the transversely isotropic media in terms of five independent stiffnesses, C_{11}, C_{13}, C_{33}, C_{44}, and C_{66}, as follows:

$$\begin{bmatrix} C_{11} & C_{11} - 2C_{66} & C_{13} & 0 & 0 & 0 \\ C_{11} - 2C_{66} & C_{11} & C_{13} & 0 & 0 & 0 \\ C_{13} & C_{13} & C_{33} & 0 & 0 & 0 \\ 0 & 0 & 0 & C_{44} & 0 & 0 \\ 0 & 0 & 0 & 0 & C_{44} & 0 \\ 0 & 0 & 0 & 0 & 0 & C_{66} \end{bmatrix}$$

Leon Thomsen's classic article on TI media (Thomsen 1986) presented a practical way of measuring and incorporating weak anisotropy into seismic imaging. He formulated a set of dimensionless parameters γ, δ, and ε (the Thomsen parameters) that yield simplified equations for P- and S-wave speeds travelling at any angle with respect to the symmetry axis:

$$\gamma = \frac{C_{66} - C_{44}}{2C_{44}}, \quad \delta = \frac{(C_{13} + C_{44})^2 - (C_{33} - C_{44})^2}{2C_{33}(C_{33} - C_{44})}, \quad \varepsilon = \frac{C_{11} - C_{33}}{2C_{33}}$$

All three parameters go to zero for isotropic materials, so you can think of the isotropy as being a special case of this more general anisotropic one. TI media have three wave modes: a P wave and two S waves — one polarized parallel to the axis of symmetry, called the SV wave, one perpendicular, called the SH. The Thomsen parameters allow us to calculate the phase velocities at any direction θ as long as the P- and S-wave velocities parallel to the axis of symmetry, denoted here as $V_{P,0}$ and $V_{S,0}$, are known:

$$V_P(\theta) = V_{P,0}(1 + \delta \sin^2\theta \cos^2\theta + \varepsilon \sin^4\theta)$$

$$V_{SV}(\theta) = V_{S,0}\left(1 + \frac{V_{P,0}^2}{V_{S,0}^2}(\varepsilon - \delta)\sin^2\theta \cos^2\theta\right)$$

$$V_{SH}(\theta) = V_{S,0}(1 + \gamma\sin^2\theta)$$

VTI anisotropy: horizontal layers. Horizontal formations with intrinsic anisotropy, like shale, as well as finely layered isotropic beds, result in transverse isotropy with a vertical axis of symmetry. The effect on a common-midpoint gather is the so-called hockey stick inflection: rays at longer offsets travelling faster than normal moveout. If the layers are dipping, and the axis of symmetry is is tilted, we call it tilted transverse isotropy (TTI).

HTI anisotropy: parallel vertical fractures and cracks. If the otherwise isotropic earth contains vertical fractures that are parallel, then there's a horizontal axis of symmetry. In this case, travel times vary as a function of azimuth. Waves go faster when they travel parallel to the fractures, slower when they travel across. HTI shows up as sinusoidal variations in travel time on a common-offset/common-azimuth seismic gather.

Orthorhombic anisotrop. A combination of HTI and VTI media occurs by having parallel vertical cracks (HTI) in an otherwise VTI medium. The nine independent stiffnesses for this system are C_{11}, C_{22}, C_{33}, C_{12}, C_{13}, C_{23}, C_{44}, C_{55}, C_{66}:

$$\begin{bmatrix} C_{11} & C_{12} & C_{13} & 0 & 0 & 0 \\ C_{12} & C_{22} & C_{23} & 0 & 0 & 0 \\ C_{13} & C_{23} & C_{33} & 0 & 0 & 0 \\ 0 & 0 & 0 & C_{44} & 0 & 0 \\ 0 & 0 & 0 & 0 & C_{55} & 0 \\ 0 & 0 & 0 & 0 & 0 & C_{66} \end{bmatrix}$$

Since there are more independent variables in this stiffness matrix, the wave speed equations become considerably more complex. But Tsvankin published a set of parameters similar to Thomsen's that reduces the number of parameters responsible for P-wave velocities in orthorhombic media (Tsvankin 1997).

The field of seismic anisotropy emerged from the need to make better seismic images by correcting velocities in the presence of anisotropy. It is also increasingly important to estimate anisotropic parameters in their own right — through seismic inversion — to serve in lithology discrimination, fracture prediction, or frackability estimation, and other reservoir characterization goals.

References

Thomsen, L (1986). Weak elastic anisotropy. *Geophysics* **51**, 1954–1966. DOI: 10.1190/1.1442051.

Tsvankin, I (1997). Anisotropic parameters and P-wave velocity for orthorhombic media. *Geophysics* **62**, 1292–1309. DOI: 10.1190/1.1444231.

Enjoy your rock and role

Futoshi Tsuneyama

Rocks are not random products of the earth. Each rock has specific character in texture, colour, odour, and even taste. Its character reflects the accumulation of its life experience. For example, sandstone shows grain-pack texture and is composed predominantly of quartz, feldspar, and lithic grains. On the other hand, shale has a finely laminated bedding texture because the main component is sheet silicate minerals, which are aligned as a result of deposition from suspension. Limestone has a micro-texture that resembles Swiss cheese: a rigid matrix with holes. Geologists describe these rock characteristics in hand specimens at the millimetre scale. Geophysicists use remote sensing to determine the rocks in the subsurface at the metre scale. Rock physics bridges between geology and geophysics.

As with the qualitative characteristics of rocks, other properties — such as seismic velocity — also vary systematically. Indeed, each rock occupies a specific territory in a crossplot of physical properties. Its territory is based on the composition of the rock but, even with the same constituents, a different texture gives us a different trend in the crossplots. Then, once we fix the constituents and texture of a rock, fluid substitution shifts the data again in the crossplots. Crucially, rock physics provides us a guide on how to look at these crossplots by simulating all of these variations.

Rock physicists start their diagnosis in the V_P versus porosity crossplot. We plot a pure-solid velocity at the zero-porosity end and a fluid velocity at the maximum porosity end. In between the fixed two endpoints, we know that the Voigt average — a convex-upward curve — makes the upper bound. It represents a pillar-type texture, which rarely exist in nature. Conversely, the Reuss average — a concave-upward curve — is the lower bound, and represents a fluid-support texture, which turns out to be a horizontal stack of solid plates with interlayer fluid in nature. Swiss cheese and grain-pack textures locate somewhere in between the two bounds; Swiss cheese has higher velocity than grain-pack, due to the higher stiffness of oval-shaped pores than inter-granular convex-surface pores. Note that Swiss cheese shows a slightly convex-upward curve, whereas grain-pack shows a much straighter trend for both V_P and V_S as a function of porosity. The shale trend stays at the bottom because of the slow velocity of clay minerals, and the laminar texture which mimics the Reuss average.

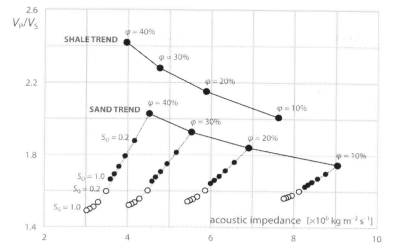

Now look at the V_p to V_s domain. Each rock has a different solid point depending on the dominant minerals:

Quartz	$V_P = 6.0$ km/s	$V_S = 4.1$ km/s	$V_P/V_S = 1.46$
Clay	$V_P = 3.8$ km/s	$V_S = 1.9$ km/s	$V_P/V_S = 2.0$
Calcite	$V_P = 6.6$ km/s	$V_S = 3.4$ km/s	$V_P/V_S = 1.94$

In a V_p to V_s plot, the Voigt bound comes to the lower side as a concave-downward curve, so we do the same for Swiss cheese. This may explain why the limestone trend is best modelled by a quadratic formula in this domain. Sandstone generates an almost linear trend and the shale trend is above that. Tsuneyama et al. (2000) reported that grainstone facies of limestone demonstrates a linear trend in this domain. Trend curvature in this crossplot indicates pore shape in the rock.

The final destination is the rock physics template (Ødegaard and Avseth 2004). Integrating knowledge of rock physics and the results of laboratory data, you can map the territories of different rocks saturated with different fluids. Then, you can apply the map in reverse to extract hydrocarbon saturation information from seismic data. The key is to do rock physics modelling first, utilizing nearby well-log data, then conduct seismic processing and analysis to reconstruct the modelling result.

This is the job of the rock physicist: enjoy your rock and your role!

References

Ødegaard, E and P Avseth (2004). Well log and seismic data analysis using using rock physics templates. *First Break*, **22** (10), 37–43.

Tsuneyama, F, I Takahashi, A Nishida and H Okamura (2003). V$_P$/V$_S$ Ratio as a Rock Frame Indicator for a Carbonate Reservoir. *First Break*, **21** (7), 53-58

Five questions to ask the petrophysicist

Chris Skelt

The distinctions between geophysics, rock physics, and petrophysics are blurred, but data transfer between the disciplines is predominantly one way. Very few petrophysicists are familiar with the realities of working with seismic data, while all geophysicists use well data prepared by a petrophysicist. This prompts the question 'are geophysics-ready packages of well data really fit for purpose?'

Here are some questions intended to stimulate discussion. They were formulated for conventional sand–shale reservoirs, but are adaptable to carbonate and unconventional reservoirs. These questions are all related to the interpretation strategy and its relevance to the fundamental subsurface building blocks. Questions in the following essay, *Five more questions to ask the petrophysicist*, deal with matters related to interpretation integrity and data quality.

1. **What does the 'shale volume' curve represent?** Ensure that the petrophysical interpretation recognizes the principal building blocks of the reservoir. Suppose a simple description of the reservoir, supported by inspection of cores, is 'alternating well-differentiated sand and shale beds with significant pore-filling clay in the sands'. Because the effects of fluid changes are localized in and governed by the inherent properties of the sands, rolling up the pore-filling clay and shale beds into a single clay or shale misses the point because it does not reproduce the key components of the reservoir. It would not permit meaningful modelling of the effects of fluid changes on elastic properties, and nor would it support a meaningful static or dynamic field model. In this case the petrophysical model needs to distinguish the sand and shale beds and quantify the clay mineral in the sands.

2. **Does the interpretation capture the appropriate level of complexity in non-reservoir zones?** Having determined that the petrophysical interpretation of the reservoir is robust, it is also necessary to focus on the non-reservoir intervals — understandably a lower priority for a petrophysicist, but important because elastic property variation in shale can be as wide as in sand and can strongly influence the seismic response. For example, in deep-water turbidite sequences shale from turbiditic and hemipelagic deposition may appear as two distinctive shale facies or a continuum of properties. And in fluvial sequences, soft, organic-rich shales deposited in swamps may

coexist with harder silty and clay-rich shales. Therefore, the petrophysical analysis should consider the lithology, properties, and distribution of non-reservoir as well as reservoir.

3. **Do the logs resolve the beds?** Many deep-water and shallow-marine reservoirs are made up of thin alternating sand and shale beds that standard logs do not resolve. If the individual beds are not resolved, it should still be possible to quantify the relative thickness of sand and shale beds within the resolution of the standard logs by reference to core images, for example. These circumstances call for a laminated petrophysical analysis quantifying the abundance and inherent properties of the sand laminations.

4. **How consistent are the interpretations from well to well?** This is obviously desirable but cannot be assumed when well data are from diverse asset groups or not evaluated simultaneously. Were the same petrophysical model and parameters used, or was each well treated as a new problem? Were all wells interpreted using the same shale trends for density or total porosity versus depth below datum?

5. **How is total and effective porosity defined?** The rock physics community is divided on whether to work with total or effective porosity. The two converge in clean sands, but each has strengths and weaknesses elsewhere. Some practitioners consider effective porosity the free fluid — the porosity that contributes to permeability and is available for hydrocarbon emplacement and production — while others use total porosity minus the electrostatically bound water. The latter may be closer to Gassmann's counsel but is abstract and awkward to invoke when constructing a petrophysical or rock physics model. Either way, clarity is essential.

Continued on page 44.

Five more questions to ask the petrophysicist

Chris Skelt

These five questions address the integrity of the petrophysical interpretation and data quality (see also previous essay).

1. **Is low saturation or residual gas present?** This may be caused by a recent (in geological time) top seal breach or leakage through a fault. It is evident from a combination of resistivity slightly higher than for brine fill with a compressional:shear velocity ratio typical of gas. Resistivity-based saturation equations do a poor job of delivering a stable result in these circumstances so the petrophysical interpretation usually needs guidance. Low saturation gas intervals can often be characterized by a steady gas saturation of about 0.1 or 0.15, but sometimes the gas saturation profile appears to taper to zero over several tens of metres. This is not a low permeability phenomenon as it occurs in formations where saturation builds rapidly at the base of the reservoir. The cause is not fully understood and may be related to diffusion occurring after reservoir leakage.

2. **Are the saturations reasonable?** Does the water saturation curve represent the petrophysicist's interpretation? Check that there is no oil or gas below the contact, or in shales — computer programs suffer from limitations of the measurements and models. Confidently modelling fluid effects on elastic properties presupposes a trustworthy interpretation of in-situ fluids as brine, low saturation gas, or a hydrocarbon reservoir with saturation defined by the equilibrium between buoyancy and capillary forces. Precise saturation is less important than this basic classification. Are saturations meaningful in the poorer quality rock where saturation equations don't work well, and computed saturation is quite dependent on choice of saturation equation? And, in view of the previous question, can the petrophysicist offer guidance on whether the initial saturations for fluid substitution should be from the invaded or undisturbed zone?

3. **Are the sonic and density logs compensated for invasion?** Invasion, particularly in gas reservoirs, affects log responses and its effects need to be removed before attempting well ties. The most transparent way of doing this is for the petrophysical analysis to quantify the three invaded zone fluids — connate water, oil or aqueous filtrate, and native hydrocarbon. If a low gas:oil

Confidently modelling fluid effects on elastic properties presupposes a trustworthy interpretation of in-situ fluids.

ratio reservoir is invaded by oil filtrate with similar properties it is unrealistic to differentiate these fluids, but nor does it matter. When working with gas reservoirs the filtrate–gas property contrast makes for a robust quantification of the invaded zone saturations. There is a related discussion to be had on the radius of investigation of the logs versus the depth of invasion. For example, shear slowness derived from the flexural wave is a shallower measurement than a direct shear or compressional arrival.

4. **Is anisotropy affecting the sonic logs?** If there is a range of relative dip (the angle between the bedding and a plane normal to the wellbore axis) textural anisotropy effects are expected in shales and thinly bedded sand–shale sequences. In the simple case of VTI anisotropy (i.e. no variation in stiffness with azimuth) increasing relative dip is associated with increasing compressional velocity and shear splitting. Algorithms to invert for anisotropy parameters and deliver sonic logs corresponding to the vertical case are becoming more available. Massive sands are usually close to isotropic, although unconsolidated sands may exhibit stress-induced anisotropy where the horizontal stiffness is less than vertical. In vertical wells it is important to test for azimuthal anisotropy with cross-dipole acquisition and processing. If azimuthal anisotropy is present in wells then the single dipole shear slowness depends on tool orientation and it may be better to estimate the shear log.

5. **Finally, what about data quality?** Are there intervals where the log readings are suspect, and does the data package include any synthetic data? If so, how were the logs synthesized? Contact theory models for dry frame bulk and shear modulus followed by Gassmann-based fluid substitution to desired saturations work well for geophysical applications. Predictions using implicit relations or neural networks are not so good at modelling the distinctive compressional and shear velocity fluid responses. And it is particularly important to distinguish synthetic from real data because the diagnostic capability of synthetic data is very limited. A simple data quality classification scheme, even if subjective, helps prevent poor quality input data and modelled results introducing bogus heterogeneity into the subsurface models.

Hard rock is (not) like soft rock

Gilles Bellefleur

In sedimentary basins rock physics is intimately associated with pore spaces and the fluids they contain. It is precisely the need to predict fluids prior to drilling that led to the development of quantitative seismic exploration tools such as amplitude variation with offset, elastic inversion, and full-waveform inversion, the interpretation of which relies on rock physics. Although not perfect, quantitative approaches provide a means of making a more informed decision before drilling, reducing the risk of a dry well.

But what about crystalline rocks? Pore space in most crystalline rocks is often negligible and typically only contains water. Therefore, fluid substitution techniques developed for soft rocks are of no or very limited use. Because of their low porosity, crystalline rocks are also significantly less affected by compaction. The general density and velocity trends in the first kilometre of well-logs acquired in older crystalline rocks are often adequately approximated with constant values. All these differences make the analysis of properties of crystalline rocks and their effects on seismic data look simpler. In a way, this is a good thing because the main difficulty for rock physics and seismic methods in crystalline rocks arises from a rather complex and uncooperative geology, often highly heterogeneous, lacking continuity, poly-deformed, and sometimes metamorphosed to the point that the original rocks are unrecognizable. It is precisely in such environments that many base-metal ore bodies are found.

In crystalline rocks, the mineralogy of the rocks is the main factor controlling the physical rock properties. Fortunately, the properties of ore-forming minerals and host rocks are well-known. Crystalline rocks hosting base-metal deposits follow the Nafe–Drake density–velocity relationship with mafic rocks, generally composed of heavy minerals having faster velocities located near the high acoustic impedance end of that relationship (Salisbury et al. 2003). Because of their high densities, sulphide minerals do not follow the Nafe–Drake relationship. Sulphide minerals have, however, a range of velocities and the expected increase in acoustic impedance due to the higher density is sometimes cancelled with lower velocity. For example, ore containing mostly sphalerite (zinc) or chalcopyrite (copper) have high density but low velocity and may produce only weak or no reflections when juxtaposed against almost any host rocks. In contrast, the acoustic impedance of a sulphide lens comprising mostly pyrite (high density and high velocity) will be significantly higher than almost any host rock. Depending on its size and shape, this lens will appear as strong amplitude reflection and/or diffraction in seismic data. Though non-economic itself, pyrite is regularly associated with economic minerals.

While bright spots on seismic data might indicate pyrite bodies and hopefully economic minerals, mafic rocks embedded in felsic rocks are a far more common explanation for high amplitude anomalies. Reflections and diffractions interpreted as potential mineral accumulations are commonly explained by small bodies of mafic rocks once boreholes intersect them. This is because ore systems are also complex and combine a variety of economic and sub-economic minerals in concentrations that can also vary spatially. Evaluating the signature of the assemblage rather than the response of individual minerals is particularly important for seismic exploration but difficult to do without a significant amount of *a priori* information. The many unknowns include composition of ore zones, their shape and size, and the composition of host rock geology. These additional complications limit the use of classic tools developed for oil and gas. This is why the seismic exploration for mineral deposits in crystalline rocks is not well-integrated with petrophysics. It also explains why physical rock properties are mostly used in a qualitative sense, providing some plausible explanation for reflections. Of course, the benefits of identifying key lithological contacts and faults cannot be neglected, but detecting ore and ultimately predicting its composition is far more appealing.

The path towards further risk reduction in crystalline rocks includes improving our understanding of wavefields produced from 3D petrophysical models representing complex but realistic ore systems. Some modelling studies suggest already that risk reduction in crystalline rocks should include mode-converted waves and 3-component receivers (Malehmir et al. 2012). The oil and gas industry taught us that quantitative approaches relying on petrophysics are the way to go. Now we must improve them for mineral exploration.

References

Malehmir, A, R Durrheim, G Bellefleur, M Urosevic, C Juhlin, D J White, B Milkereit, and G Campbell (2012). Seismic methods in mineral exploration and mine planning: A general overview of past and present case histories and a look into the future. *Geophysics* **77**, WC173–WC190. DOI: 10.1190/geo2012-0028.1.

Salisbury, M H, C Harvey, and L Matthews (2003). The acoustic properties of ores and host rocks in hardrock terranes. In: Eaton, D , B Milkereit, and M Salisbury, eds., *Hardrock seismic exploration*. Tulsa, OK, Society of Exploration Geophysicists, 9–19.

Here be dragons: the need for uncertainty

Franck Delbecq

We should regard uncertainty as an essential deliverable accompanying any result we present.

Quantitative interpretation (QI) — seismic inversion together with rock physics analysis — is a vehicle to express seismic data in terms of petrophysical parameters. The emergence of QI has brought new tools to analyze and communicate geophysical data in geological, geomechanical, and engineering terms.

But — there is always a 'but' — we often present QI results with no mention of the associated uncertainty, error bars, or assumptions. We know that all results have uncertainties, and yet the moment we display these results in the form of a map, graph, or 3D volume, the associated uncertainty seems to be forgotten. That is a disservice to the data, at best, and perilous at worst. Instead, we should create 'maps of ignorance', letting everyone know about pitfalls: where the dragons are.

An extract from Olaus Magnus's 1539 Carta Marina. *Mapmakers were drawing dragons and sea monsters beyond the edges of the known world to warn travellers that they were entering less reliably mapped territory.*

There are many types and sources of uncertainty in a QI workflow:

- **The input data.** Are the acquisition and processing reliable? Are they consistent throughout the survey?
- **The lack of data locations.** Do I have enough wells for a reliable initial model?
- **Unrepresentative sampling.** Do the wells over-represent anomalous locales?
- **The processing/inversion parameters.** Wavelets, inversion constraints, V_p/V_s background, and so on.
- **The model or assumptions used.** Is it okay to assume an isotropic world here?
- **The algorithm.** Is a two-term linear AVO approximation suitable for my data?
- **The non-uniqueness of the inverse model.** More than one combination of elastic properties can fit the seismic expression. More than one rock type can have these elastic properties! (But some are more probable than others.)
- **The uncertainty in the calculated uncertainty itself.** Were all the main drivers captured when calculating the confidence level? Do I grasp the significance of various effects?

The challenge is to come up with a way to capture how these various uncertainties combine. If nothing else a qualitative assessment will go a long way; capturing and mapping the main sources of uncertainty will help develop a sense of what is reliable or not.

One approach is to propagate the different uncertainties onto the rock physics template. The use of rock physics is critical because it assesses the effect of different uncertainties on what matters to a specific play (porosity, TOC, or facies, for example). Using crossplots, rock physics templates, and probability distribution functions can provide a comprehensive, quantified, and visual perspective on uncertainty (Moyen 2013 and Nieto at al. 2013).

Stating uncertainty will not cast doubt on our work. In fact the opposite is true! Communicating the uncertainty imparts greater information and, therefore, confidence. In some cases the uncertainty can actually be directly used as an input of the next step, such as a constraint in geomodelling (Thenin et al. 2013) — another strong reason to present the quantified uncertainty associated with our models.

References

Moyen, R (2013). Using uncertainty in quantitative seismic characterization. *CSEG Recorder* **38** (9), 24–29.

Nieto, J, B Batlai, and F Delbecq (2013). Seismic Lithology Prediction: A Montney Shale Gas Case Study. *CSEG Recorder* **38** (2), 34–41.

Thenin, D and R Larson (2013). Quantitative seismic interpretation — an earth modeling perspective. *CSEG Recorder* **38** (9), 30–35.

How to catch a shear wave

Arthur Cheng

Acoustic logging is the primary method for measuring the elastic properties of rocks in situ. But measuring the speed of waves, especially shear waves, in rocks is not straightforward. End-user geophysicists typically get just the compressional and shear slowness (reciprocal of velocity) from service companies, with little or no indication of how robust those measurements are. I want to tell you how those data are obtained, and what to watch out for.

Some common applications of acoustic logging data are:

- Elastic properties and reservoir characteristics, rock physics modelling, and fluid substitution.
- Synthetic seismograms from compressional and shear slowness.
- Permeability and fracture detection from Stoneley waves.
- Azimuthal shear wave anisotropy from cross-dipole measurement.
- In-situ stress inferred from azimuthal shear wave anisotropy.
- Near borehole reflections.

Modern acoustic logs measure compressional and shear wave slowness parallel to the borehole, which means that in a vertical borehole they measure the vertical slowness. In addition, the dipole log provides directional shear slowness.

Wave types

There are two types of waves generated by acoustic logging tools: body waves and guided waves. The body waves propagate as a compressional wave in the borehole fluid and are transmitted into the formation. A portion is critically refracted, travels as a head wave along the borehole–formation interface, and refracts back into the borehole fluid where it is detected by the receiver array.

Guided waves are surface waves guided by the borehole itself. They include the pseudo-Rayleigh wave, Stoneley wave, and flexural wave (also known as the dipole mode wave). These waves are affected by changes in the properties immediately surrounding the borehole such as casing and invaded or damaged zones.

When you consider the variety of wave modes around an acoustic tool,
it seems remarkable that we can extract such useful data as we do.

EMPIRICAL • PETROPHYSICS • ANISOTROPY

Shear slowness

One factor affecting the measurement of shear wave slowness is the slowness of the compressional wave in the borehole fluid. When the fluid slowness is less than the shear slowness, there cannot be a refracted shear wave arrival. Such a formation is known as a 'slow' or 'soft' formation whereas a 'fast' or 'hard' formation is one where there is a refracted shear. Notice that the distinction is based on the borehole fluid slowness, which can range from 1600 m/s to over 1015 m/s, depending on the properties of the drilling fluid. A conventional acoustic logging tool that relies on a pressure source (also known as a monopole source) cannot measure shear wave slowness in a soft formation.

The one way to directly measure shear wave slowness is by the use of the flexural wave. The flexural wave is generated by a displacement source (a dipole source) in the borehole and is measured by the pressure difference on opposite sides of the tool. The flexural wave is a surface wave, similar to the Stoneley wave, the distinction being that the flexural wave has the sides of the borehole moving out of phase (one side positive, the other negative) rather than in phase. This results in no net volume change in the borehole fluid, in contrast to the Stoneley wave. The flexural wave travels at the formation shear velocity in all formations, and approaches the Stoneley wave slowness at high frequencies. Because the dipole source is directional, one can measure azimuthal shear wave anisotropy with two orthogonal dipole sources, also known as cross-dipole source.

When you consider the variety of wave modes around an acoustic tool, it seems remarkable that we can extract such useful data as we do. This is the miracle of geophysics!

Mapping fractures

Peter Duncan

Many applications of rock physics are directed at understanding how rocks will fail in order to prevent failures. But since oil and gas production has moved into rocks with nanodarcy permeability, engineering rock failure by the process of hydraulic fracturing is necessary.

Hydraulic fracture stimulation was developed in the 1940s by Standard Oil of Indiana and licensed to Halliburton in 1949. Since then it has been applied to more than 2.5 million wells to stimulate production by increasing the effective drainage volume of the treated wells. For the most part treatments are designed with simplistic radially symmetric (1D) models that predict a bi-wing symmetric, penny-shaped tensile fracture, opening in the direction of least horizontal stress. These models are supported by laboratory experiments using homogeneous isotropic rock samples. But the real world is much more complex. Mine-back experiments on shallow hydraulic fracture treatments performed by Norman Warpinski and others at Sandia Laboratories at the Nevada Test Site in the late 1970s and early 1980s demonstrated that rock fabric, inhomogeneities, joints, and pre-existing fractures control the fracture morphology. More recently, microseismic monitoring has shown that the bi-wing symmetric penny shaped fracture rarely, if ever, occurs.

In principle, hydraulic fracturing works by increasing the pore pressure in the rock, which decreases the net effective confining pressure. Consider the conventional Mohr's circle plot opposite. The effect of increasing the pore pressure is to move the circle to the left. When it intersects the so-called 'line of friction' with slope μ then shear failure begins. When σ_3 becomes negative, tensile failure can occur. The stress conditions for failure can be different for specific joint sets, faults, or rock fabric, thus resulting in different kinds of failure at different times.

The rupture of the rock is usually accompanied by the release of seismic energy. The magnitude and frequency content of the emitted sound is related to the size of the failure plane. The seismic signal also exhibits a radiation pattern that depends upon the type of failure — tensile, strike slip, dip slip, etc. The aim of microseismic monitoring is to characterize the rock failure as completely as possible by analyzing the seismic signal it created.

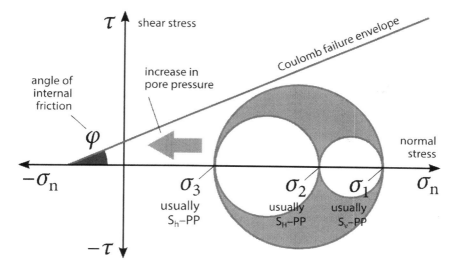

The location of the rupture in time and space (the hypocentre) can be determined from the arrival time of the signal at one or more receiver stations. Estimation of the hypocentre from a single station requires accurate picks of the arrival times of the compressional (P) phase and shear (S) waves, the tilt of the wavefield in three dimensions, and an estimate of the seismic velocity of the earth. Observations from multiple stations surrounding the event location can relax the need for accurate picking and increase the accuracy of the hypocentre estimate. A beamforming or full waveform migration approach to hypocentre location is also possible. This eliminates the time-consuming and often inaccurate travel-time picking process. The accuracy of this technique increases as the temporal and spatial wavefield sampling increases.

If the wavefield sampling is sufficient to describe the radiation pattern, then we can solve for the failure mechanism and the local stress state at the point of failure. The mechanics of the failure are described by a tensor of rank three, or moment tensor. Symmetry dictates that there are only six independent terms in the tensor and, in the case of pure shear failure, the number of independent terms reduces to three. A full moment tensor inversion provides an estimate of each of the independent moment terms and the magnitude of the moment that created the event. Knowing these values we can make an estimate of the size, strike, dip, and rake of the fault segment that moved to create the event. From the population of all events, geoscientists can build a discrete fracture network that is representative of the stimulation, insofar as the recovered acoustic signal is representative of the rock that broke. This network represents the enhanced permeability pathways from which the well production can be estimated.

Mathematical descriptions of physical phenomena

Rocky Detomo

Most geoscientists find themselves in a situation where they are trying to relate different rock property measurements from many different wells and understand the relationship between these properties. Some rock properties are relatively straightforward to understand and describe, e.g. the density of a 'rock', ρ, is a volumetric average of its component masses:

$$\rho = \frac{m_f + m_1 + m_2 + \ldots + m_n}{v_f + v_1 + v_2 + \ldots + v_n}$$

where m_i are the masses of the n component rock materials, v_i are their volumes, and v_f and m_f is the saturating fluid's mass. For a clastic, three-component system (sand, shale, brine) this is simply expressed as:

$$\rho = (1 - \varphi)(N \rho_{sa} + (1 - N) \rho_{sh}) + \varphi \rho_f$$

where φ is the rock's porosity, N is the sand fraction of the sand–shale mix, and ρ_{sa}, ρ_{sh} and ρ_f are the densities of the sand and shale grains, and the fluid, respectively.

Other rock properties are more complex to describe (such as rock velocity, V), since this rock property is the result of chemical and physical processes that have taken place over long periods of geological time. Therefore, the velocity of a rock cannot necessarily be calculated directly from its contributing components, although if these historical processes can be understood and estimated, some prediction and/or bounds on these values can often be made.

In addition to relating rock properties of a particular rock to each other, we often want to describe how these rock properties are changing over time. Over geologic-scale time intervals, rocks undergo pressure changes (e.g. compaction with burial depth, overpressures etc.), thermal changes (temperature changes with depth and proximity to thermal conduits or salt), and chemical changes (carbonate dissolution, diagenesis, quartz cementation, etc.). We often attempt to understand these temporal effects by relating how the properties of rocks of similar composition vary as a function of burial depth. Therefore, it is common to see plots such as the one opposite, which are assembled from many well-log measurements of compositionally similar materials.

It is most common to fit these data to some mathematical function, such as a second-order polynomial, as shown below. However, there is often no natural or physical basis for such a choice in functional form, and these choices often display behaviour outside of the input data's range that are completely unphysical. In addition, the variance in the data is often attributed to 'natural variability', and is simply carried as a fitting uncertainty.

However, if we examine the data more closely, and we account for the asymptotic values for shale density both at the surface and at great depth, we might choose an exponential fit for the functional form instead, allowing us to examine the remaining variability in terms of other rock properties and burial history. Note that in the figure below although the polynomial seems to describe the data well enough in the range of measurements, its value both at the surface and at great depth is both unphysical and unrealizable, leading us to believe that a polynomial fit is a poor choice.

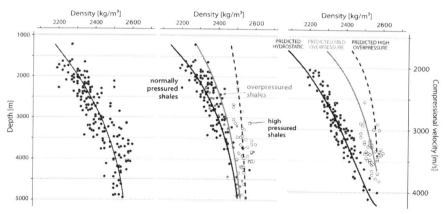

On the left, a typical crossplot of shale density versus burial depth with a best-fit arbitrary choice of a second-order polynomial fit. Classification of these shale densities in terms of their pressure regime gives the middle plot, where some of the variability is clearly associated with the shale's pressure history. This gives rise to a calculation of an exponential function form with depth whose coefficients are pressure history dependent, as shown by the three pressure curves displayed. The right plot shows the resultant shale velocity crossplotted with shale density, where remaining variability is associated with variations in silt content.

Classification of the shale densities in terms of their pressure regime gives the centre plot, where some of the variability is clearly associated with the shale's pressure history. Close examination of other log data rock properties allows us to characterize much of the remaining variance as a result of the shale's relative silt content. This more holistic description gives insights into what is driving the vertical and lateral viability in reflectivity that results from these rock properties. Recognizing and describing the impact of pressure on shale densities gives a more consistent description of other rock properties. This description with respect to a rock's burial history is fundamental to understanding seismic reflectivity, and rock property models should be tied more closely to their associated basin model.

Measurements are scale dependent

Sagnik Dasgupta

Minerals have a range of crystal symmetries, from isotropic to triclinic. Until the last century, however, sedimentary rocks were modelled as isotropic media. How reasonable is this assumption?

To answer this question, we need to understand the characteristic length of different elastic stiffness experiments. Currently, the measurements vary from nano-indentation, ultrasonic transmission, and borehole seismic to surface seismic amplitude and seismic travel-time inversion. Nano-indentation is an extremely high-resolution measurement from which we can measure the elasticity of a single mineral or grain. The other measurements are all forms of elastic wave propagation experiments, varying in frequency from 800 kHz for ultrasonic down to about 2 Hz for seismic travel-time measurements.

Another important concept is the representative elementary volume (REV) of the medium, defined by Hill (1963) as 'the smallest volume over which a measurement can be made that will yield a value representative of the whole sample.' The nano-indentation experiment has an extremely small volume of investigation (VOI) — smaller than the REV of the material. On the other hand, it helps in understanding the properties of individual constituents, which helps to build the composite for the REV. For all the other experiments, the VOI is inversely related to the frequency of the measurement and is generally larger than the REV of the medium.

Measurement	log(resolution) [m]	log(VOI) [m³]
Nano-indentation	−5.2	−20
Ultrasonic	−1.3	−4.6
Sonic	+0.26	−0.26
Borehole seismic	+1.0	+4.5
Seismic amplitude	+1.3	+6.7
Seismic velocity	+2.8	+9.6

Notice the enormous variance in the VOI for the various methods. For example, a mineral crystal of 1 mm³ (i.e. 10^{-9} m³) is larger than the VOI for nano-indentation, while the VOI corresponding to seismic travel-time measurements is larger than 1 km^3 (i.e. 10^9 m^3) and encompasses a significant number of sedimentary layers.

Above the REV we have the notion of an effective medium approximation (EMA) in which the medium is considered to be a continuum and statistically uniform. The elastic properties of the EMA can be exactly described using the precise description and statistical distribution of the individual constituents. Well-known attempts have been made to describe such mixtures by Voigt (1890), Reuss (1929), Eshelby (1957), Hashin and Shtrikman (1962), and Backus (1962).

The elastic properties of the effective medium are scale dependent. Furthermore, heterogeneity (property variation with position) and anisotropy (property variation with direction) are also scale dependent. For example, a material can be homogeneous and anisotropic on the large scale and heterogeneous and isotropic on the small scale. The effective medium concept allows us to understand and reconcile measurement variations across very different scales.

References

Backus, G E (1962). Long-wave elastic anisotropy produced by horizontal layering, *Journal of Geophysical Research* **67** (11), 4427–4440. DOI: 10.1029/JZ067i011p04427.

Eshelby, J D (1957). The determination of the elastic field of an ellipsoidal inclusion, and related problems. *Proceedings of the Royal Society of London*, Series A **241**, 376–396. DOI: 10.1098/rspa.1957.0133.

Hashin, Z and Shtrikman, S (1962). A variational approach to the theory of the effective magnetic permeability of multiphase materials. *Journal of Applied Physics* **33**, 3125–3131. DOI: 10.1063/1.1728579.

Hill, R (1963). Elastic properties of reinforced solids: some theoretical principles. *Journal of the Mechanics and Physics of Solids* **11**, 357–372. DOI: 10.1016/0022-5096(63)90036-x.

Reuss, A (1929). Berechnung der Fließgrenze von Mischkristallen auf Grund der Plastizitätsbedingung für Einkristalle . *Zeitschrift für Angewandte Mathematik und Mechanik* **9**, 49–58. DOI: 10.1002/zamm.19290090104

Voigt, W (1890). General theory of the piezo- and pyroelectric properties of crystals. *Abh Gött* **36** (1), 1–99.

Meet Gassmann and Biot

Matt Hall

A few of the names we hear in geophysics — Snell, Gauss, Euler, Maxwell — belong to lauded scientists whose life stories are well-known. Others from the modern era, for example Leon Thomsen and Guus Berkhout, are familiar faces and indeed you may have met these individuals. Still others, however, such as Karl Bernhard Zoeppritz, Keiiti Aki, and Paul Richards, are familiar to the theorists, but the details of their lives are not widely known. Did you know that Zoeppritz died at the age of only 26? Or that Aki devised the moment magnitude scale that replaced the Richter scale? Or that Richards teaches an undergraduate course at Columbia called Weapons of Mass Destruction?

A few names come up a lot in rock physics, usually in pairs: Gassmann and Biot, Hashin and Shtrikman, Reuss and Voigt, Young, Lamé, and so on. Although I've written previously about Young (Hall 2014), and Bill Goodway touched on Lamé before (Goodway 2012), I wanted to know more about these people — were they geoscientists? Where did they work? Are they still alive?

Let's start with the two people who had the biggest influence on how we treat the effects of fluids on rocks: Fritz Gassmann and Maurice Biot.

Fritz Gassmann (1899–1990)

One of the best-known names in rock physics, Gassmann was born in Zürich at the turn of the 20th century, and spent most of his career there. As a young man at ETH Zürich, Gassmann was a student of George Pólya, the eminent mathematician who finished his career at Stanford and is maybe best known for the book *How to Solve It* (Doubleday, 1957). Gassmann's early work focused on group theory, and he obtained some important results in that field before moving into geophysics in about 1928. For the next 15 years or so he was exceptionally busy — he worked at the Swiss Seismological Service and taught mathematics at a high school, all while lecturing in geophysics at ETH Zürich. On becoming a professor there in 1942 he founded the Institute of Geophysics. Perhaps his most important contributions were *On the elasticity of porous media* and *Elastic waves through a packing of spheres*. These were the first works in geophysics to treat rocks as porous materials and are both still cited today. Gassmann retired in 1969, and I can't find a single word written about him after that.

Gassmann

Biot

Maurice Biot (1905–1985)

Showing early signs of tenacity, Biot collected degrees in philosophy, mining engineering and electrical engineering in his native Belgium, before moving to the US in 1931. Like Gassmann, Biot had a sound mathematical foundation — his PhD supervisor at CalTech was Theodore von Kármán, the aeronautics pioneer and eponym of cloud vortices among other things. Between 1935 and 1962 Biot worked on the problem of poroelasticity — the mechanics of fluid-saturated porous media, now often called Biot theory. He published many papers on the subject, the two most prominent of which were *General theory of three-dimensional consolidation* and *Theory of propagation of elastic waves in a fluid-saturated porous solid*, which came in two parts. Between them, these papers have over 17 000 citations. An intuitive scientist, his great talent seems to have been applying his ideas across the disciplines. Quoting from his citation for honorary fellowship in the Acoustical Society of America:

> ...*in the 1930's he studied wave propagation in prestressed solids, formulated for them an elegant general system of linear field equations, applied these in geophysics, seismology, and engineering, and ultimately found a method for dealing with the early stages of folding in geological structures.*

I think that's a level of curiosity we should all aspire to!

References & acknowledgments

Biot, MA (1941). General theory of three-dimensional consolidation. *Journal of Applied Physics* **12** (2), 155–164. DOI: 10.1063/1.1712886.

Biot, MA (1956). Theory of propagation of elastic waves in a fluid-saturated porous solid. Part I: Low-frequency range and Part II: Higher-frequency range. *The Journal of the Acoustical Society of America* **28** (2). Part I: 168–178, Part II: 179–191.

Gassmann, F (1951). Über die Elastizität poröser Medien. In: *Vierteljahrsschrift der Naturforschenden Gesellschaft in Zürich.* Band 96, S 1–23. Available in English, thanks to James Berryman, at *ageo.co/gassmann51*

Gassmann, F (1951). Elastic waves through a packing of spheres. *Geophysics* **16** (4), 673–685. DOI:10.1190/1.1437718.

Goodway, B (2012). The magic of Lamé. In: *52 Things You Should Know About Geophysics*. Mahone Bay, NS: Agile Libre.

The Poromechanics Internet Resources Network has a tribute to Biot which proved very useful: *ageo.co/1U4xtSD*

The image of Gassmann is © Ammann Photo, Zürich, and licensed from ETH Bildarchiv. That of Biot is reproduced with the kind permission of Mme. Nady Biot.

Meet Hashin and Shtrikman

Matt Hall

Hashin and Shtrikman get plenty of recognition — 6 of the 52 essays in this book mention them. But, provoked by Alan Cohen's (2012) essay in *52 Things You Should Know About Geophysics*, I wanted to know more about the scientists behind the famous bounds. Since Hashin always gets top billing, let's start with Shmuel Shtrikman.

Shmuel Shtrikman (1930–2003)

Born in Brest, Poland (now Belarus), Shtrikman's family moved to Israel in 1934 when he was four years old, eventually settling in Tel Aviv. After four years of study at the Technion (also known as the Israel Institute of Technology), he joined the Department of Electronics at the Weizmann Institute of Science, a small research university, in 1954. He worked with Hashin on composite materials and their 1963 paper, *A variational approach to the elastic behaviour of multiphase material*, was a landmark in the science of materials and rock physics. This built on a precursor paper which gave the magnetic view of the same kinds of materials: *A variational approach to the theory of the effective magnetic permeability of multiphase materials*. Both papers have thousands of citations. By all accounts, Mula — as he was known — was both a vivacious character, and a precocious scientist. From an obituary by Doyle et al. (2004):

> *Mula was continually involved in the design and development of an eclectic array of novel devices, including microstrip antennas, miniature cryogenerators, eardrum vibration detectors, infrared detectors, laser drills, trace gas detectors, linear motors, torque couplers, and even a hair clipper! One of his most widely used inventions identifies gems by laser fingerprinting.*

By all accounts, Mula — as he was known — was both
a vivacious character, and a precocious scientist.

Zvi Hashin (born 1929)

Hashin's family also fled pre-war eastern Europe. He was born in the Free City of Danzig, a protectorate of the League of Nations for most of the interwar period and now Gdańsk, Poland. His family left for Israel in 1936 and settled in Haifa. After the war, Hashin served in the military for a short time, and then entered the Technion in 1950 to study mechanical engineering. However, although Hashin and Sktrikman were contemporaries at the Technion, he describes their collaboration as starting in 1961, while he was a faculty member at the University of Pennsylvania and Shtrikman was a visiting faculty at the nearby Franklin Institute Laboratories. They were working on different aspects of composite materials, with Hashin, as the mechanical engineer, more interested in elastic properties. The work was highly sought after, especially after composite materials became widespread in aviation and civil engineering applications.

In 1973, Hashin founded the Department of Mechanics, Materials and Structures at Tel Aviv University. Over the next two decades, he went on to spend time there and at several other institutions — Pennsylvania (again), Ecole Polytechnique, Berkeley, and Cambridge — as well as consulting to the US Navy and Air Force, and NASA. He retired from academia in 1997, and was awarded the Benjamin Franklin Medal in 2012, recognizing his contributions to the science of materials.

References & acknowledgments

Cohen, A (2012). Why you care about Hashin–Shtrikman bounds. In: *52 Things You Should Know About Geophysics*. Mahone Bay, NS: Agile Libre.

Doyle, B, D Treves, P Flanders, S Schultz, F Friedlaender, H Thomas (2004). "In Memoriam: Shmuel (Mula) Shtrikman (1930–2003)". *Transactions on Magnetics* (IEEE) **40** (6): 3441–3442. DOI:10.1109/TMAG.2004.837391.

Hashin, Z and S Shtrikman (1962). A variational approach to the theory of the effective magnetic permeability of multiphase materials. *Journal of Applied Physics* **33**, 3125–3131. DOI: 10.1063/1.1728579.

Hashin, Z and S Shtrikman (1963). A variational approach to the theory of the elastic behaviour of multiphase materials. *Journal of the Mechanics and Physics of Solids* **11** (2), 127–140. DOI: 10.1016/0022-5096(63)90060-7.

Hashin, Z (1980). Citation Classic: Variational approach to the theory of the elastic behavior of multiphase materials. *Physical Chemical & Earth Sciences* **6** (11) p16.

My box of rocks

Stephen Brown

One of my first memories concerning rocks, in fact my first formal introduction to geology, was at about age 10 during a family vacation to Yellowstone National Park. Before the trip my father took me to our neighbourhood western clothing store and fitted me out with a gleaming pair of cowboy boots, stiff blue jeans, and a shirt with pearl snaps. After an eternity of driving from Salt Lake, we finally arrived at a Jackson Hole motel constructed of whole pine timbers and old wagon wheels. That night we went to the town square and witnessed a mock stagecoach holdup — one of the men on the coach was shot by a bandit and fell to the ground in a heap right by my feet, his crisp white shirt stained with ketchup. I was horrified and inconsolable.

My parents took their crying child to an open-air tourist shop and bought me two things — a hematite ring engraved with a Roman gladiator, and a rock collection. The first I adored — so beautiful, so useful — the second I hated. The 20 or so small rocks and minerals were irregular, ordinary, and to top it off were glued to the bottom of their respective isolated tiny compartments — immovable, untouchable, and devoid of context. The most boring things imaginable. The ring travelled with me everywhere; the rocks collected dust in my bedroom closet.

Now here I am — a sixty-something geologist, a geophysicist, and a rock physicist. How did this happen? Was it the idea of Earth's hidden treasures that reeled me in? In a way yes, but not how you might think. I found a way to come to terms with that box of rocks.

I have spent a good amount of time seeking treasure or helping others do the same, from panning for gold and looking for trilobites with my father, later working in my father's various gold and silver mines, later still working in the oil and gas industry. This has all, in itself, been rather unsatisfactory. What has really kept me going is seeking and observing nature's beauty. That same beauty that was hidden within that box of rocks.

Upon closer inspection, those rocks had patterns and structure in their seemingly rough edges. If they were freed from their glue, I could ask them questions, and they could respond. As a child, the gold ore looked like a piece of gravel.

That same beauty that was hidden within that box of rocks.
...If they were freed from their glue, I could ask them questions,
and they could respond.

The iron pyrite was, as I knew, only fool's gold. If I had listened carefully then, as I do now, the gold ore would have spoken of ancient mineral-laden fluids, the pyrite with its proud crystal facets would be unapologetically beautiful. Understanding their beauty from the spatial structure of microcracks, porosity, and mineralogy all the way to the fit of the continents in ancient Pangea provides an order to things that is perpetually interesting, compelling, and practical.

Lately, I have been pondering two points of view. That of the craftsman, who creates an object for a single purpose of functionality or beauty, and that of the scientist, who takes what nature has provided, observes the order within, and learns how we fit in. My goal now is to strike a balance between these two.

Negative Q

Christopher Liner

Seismic Q is a measure of attenuation. To be specific, Q is inversely related to attenuation: small Q means large attenuation and vice versa. Petroleum seismology field measurements below 100 Hz strongly suggest that earth's layering leads to attenuation with linear frequency dependence, corresponding to constant Q. Higher-frequency methods, such as sonic logs and lab ultrasonic measurements, point toward viscous attenuation effects in which attenuation is a nonlinear function of frequency. But we are concerned here with the layer-induced constant Q theory of low frequency seismic waves.

In the constant Q theory, there is an exact relationship between attenuation and dispersion (frequency-dependent velocity):

$$Q = \cot\left(\pi \frac{\ln(V/V_0)}{\ln(f/f_0)}\right)$$

where V_0 is velocity observed at some reference frequency f_0, V is velocity measured at a higher frequency f, and ln is the natural logarithm. A positive Q value implies attenuation and necessarily leads to dispersion so that high-frequency waves travel faster than low-frequency waves. The opposite case, velocity decrease with increasing frequency, corresponds to negative Q. Negative Q is not observed in the lab where small rock samples are used and only intrinsic attenuation mechanisms can be tested.

To understand the source and implications of negative Q, we need to look at the relationship between earth's layering, anisotropy, and attenuation. Backus averaging is a fundamental theory that considers the interaction of low-frequency waves with a fine-layered earth model. Backus averaging says that long waves in such an earth will appear to be moving through an anisotropic earth, even if each thin layer is isotropic. Layer-induced anisotropy is well documented in field data and anisotropy corrections are routinely applied in seismic imaging. Total anisotropy is the combination of layer-induced and intrinsic anisotropy (e.g. in shales). However, from the viewpoint of data processing it makes no difference if the anisotropy is apparent or intrinsic, we simply estimate total anisotropy and apply corrections.

Backus averaging theory explains layer-induced anisotropy, yet seems silent on

Constant Q theory combines with Backus averaging to allow calculation of local Q values as a function of depth. In sonic low-velocity zones, this process implies negative Q.

the other important effect of layering — attenuation. But there is a connection and it leads to the strange possibility of local negative Q values.

Backus averaging applies a moving vertical average to the thin-layered elastic earth model provided by sonic logs. The result is a smoother, more anisotropic earth than the original.

So how does this relate to attenuation? The Backus averaging earth model is smoother than the original because of the vertical averaging operation. Consider a 20 kHz sonic log that is Backus averaged to the equivalent of 100 Hz, and further consider a low-velocity bed as observed by the sonic log. As Backus averaging progresses, the low-velocity zone will be pulled higher by velocity values above and below. Thus, the velocity seen by a 100 Hz wave will be greater than the velocity seen by a 20 kHz wave. The key point is that Backus averaging describes frequency-dependent anisotropy and velocity. The latter is apparent dispersion due to layering, no less real than apparent anisotropy due to layering.

Constant Q theory combines with Backus averaging to allow calculation of local Q values as a function of depth. In sonic low-velocity zones, this process implies negative Q — high frequencies travel slower than low frequencies.

Can this really be the case? About half of all reflection coefficients are negative, implying transmission coefficients greater than one ($R + T = 1$). The field is amplified but energy is still conserved. Perhaps negative Q is simply the multi-layer extension of this well-known phenomenon.

Further reading

Liner, C (2014). Long-wave elastic attenuation produced by horizontal layering. *The Leading Edge* **33** (6), p 634–638. DOI: 10.1190/tle33060634.1.

Pitfalls of anisotropy

Paul Anderson

The effect of velocity anisotropy on well logs is not only well documented in literature, it is exploited whenever possible in seismic data processing and rock properties analysis. We use it for the characterization of fractures in or above a reservoir, and for the optimization of parameters for prestack depth migration. Since Leon Thomsen's classic paper (Thomsen 1986), the presence of velocity anisotropy has become more than accepted, it is expected. While it is now routine to correct for anisotropy in processing, what happens when we forget to include it in more traditional rock physics analysis for AVO or prestack inversion for reservoir descriptions or modelling?

In an example from offshore northwestern Australia, a more detailed investigation revealed that the wells with the more significant change with depth were the ones drilled along deviated paths, with deviations from vertical as high as 73°. A number of other authors have described this effect and how to mitigate it (Vernik 2008; Hornby et al. 2003). Locally bedding is near horizontal, allowing for a VTI assumption to be tested to correct this effect. By iterating though various estimates of Thomson's anisotropic parameters ε and δ, a simple correction can be made to the deviated wells to correct them to a vertical velocity profile as shown in the example opposite from the above mentioned wells.

This is often what is delivered to the geophysicist by the petrophysicist in order to produce logs for AVO models. While the effect of compaction on the sonic velocity is obvious, it is not immediately obvious what the compaction trend should be. The sonic velocities for the various wells appear to be different, suggesting that perhaps a different compaction trend is present around the field. However, the wells are within a few kilometres of one another so this seems unlikely.

While this example demonstrates the simple correction that can be made to the well logs, it's also clear there are potential pitfalls. When a petrophysicist (or well-log software) provides you with well logs that have been corrected to true vertical depth, do they account for this anisotropy or do they simply resample the logs? What if no vertical wells are available for calibration? What impact would this have on your estimated rock properties? Typically most petrophysicists don't correct for this effect, simply because they haven't been asked to do so, and even if they are asked, the solution to ε and δ is often not known. One

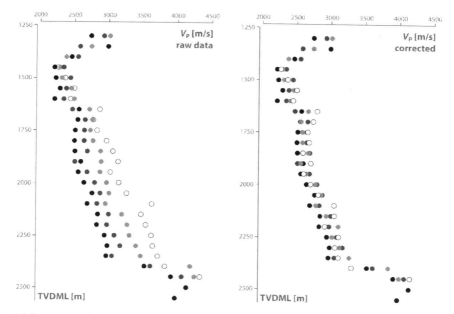

Left: P-wave sonic velocity versus vertical depth below mudline for four wells from a field on the northwest shelf of Western Australia. Note that the compaction trend for the deviated wells (white and pale grey) appears different from the vertical wells (dark grey and black). Right: The same data after correction for VTI anisotropy. The deviated wells now follow the same compaction trend as the nearby vertical wells.

begins to wonder how many rock physics and AVO studies have been done with the incorrect values for P-wave and S-wave velocity (remember anisotropy impacts S-waves, too). This would also result in anomalous drift curves when correlating to seismic data with either manual or recorded checkshot surveys. Also, if AVO models, compaction trends, and low-frequency models for poststack or prestack inversion were built using the uncorrected logs (e.g. Rowbotham 2003), it would not be hard to understand why our AVO models often do not match the real seismic results. Have we missed prospects simply because we didn't account for anisotropy in our modelling? While the correction is easy in principle, application of the correction may not be so easy due to a lack of data with which to calibrate.

References

Hornby, B, J Howie, and D Ince (2003). Anisotropy correction for deviated-well sonic logs: Application to seismic well tie. *Geophysics* **68** (2), 464-471. DOI: 10.1190/1.1567212.

Rowbotham, P, D Marion, R Eden, P Williamson, P Lamy, and P Swaby (2003). The implications of anisotropy for seismic impedance inversion. *First Break* **21** (5), 24–28.

Thomsen, L (1986). Weak elastic anisotropy. *Geophysics* **51**,1954–1966. DOI: 10.1190/1.1442051.

Vernik, L (2008). Anisotropic correction of sonic logs in wells with large relative dip. *Geophysics* **73** (1), E1–E5. DOI: 10.1190/1.2789776.

Pitfalls of the laboratory

Fuyong Yan

In the academic field of seismic rock physics, there are two classes of practitioners: those who specialize in theoretical modelling and applications, and those who specialize in laboratory experiments. Quite often these two groups of people are prejudiced against each other: the laboratory geophysicist thinks that the theoretical geophysicist only knows about playing games with equations, and does not understand the physics of real rocks; conversely the theoretical geophysicist laughs at the mathematical skills of the laboratory geophysicist. I am lucky that I participate equally in both theoretical modelling and laboratory measurements. And I believe these opposing camps need each other more than they realize.

The two classes are far from equal; there are far fewer people focusing on laboratory study. One reason might be that the lab is much less rewarding:

- It needs substantial financial investment in equipment.
- It may take years to build the measurement system.
- It requires a knowledge of electrical engineering and mechanics.
- Sometimes you need to be a handyman to make gadgets by yourself.
- You don't expect too many publications from the laboratory study.

Because of all this, laboratory data are scarce. We really should give more credit to the lab heroes. For theoretical study, one needs only a computer, and a motivated person can produce several peer-reviewed journal papers per year.

The under-appreciation of lab data is coupled with an under-appreciation of its caveats. Too often, lab data are blindly trusted and over-interpreted. It's true that unlike seismic exploration, in the laboratory, we have tight control on the objects we measure: we accurately know the dimensions of the sample and the path of wave propagation, and we can acquire other information about the rock, such as porosity, permeability, and mineral composition. So why wouldn't you trust laboratory measurement data?

Here's one example why. About 10 years ago, my lab built a velocity anisotropy measurement system following Hua Wang's idea (Wang 2002). I started to do the measurement and found a lot of negative c_{13} for the organic shale samples

I measured. What is the physical meaning of negative c_{13}? I put the question on LinkedIn's rock physics group. Many experts gave explanations. But none of them were really relevant.

For a transverse isotropic (TI) layered medium, there are three principal Poisson's ratios. One of them is defined as:

$$\nu_V = \frac{c_{13}}{2(c_{11} - c_{66})}$$

As shown, this Poisson's ratio ν_V is the ratio of lateral strain to axial strain when an axial stress parallel to the TI symmetrical axis is applied to the TI medium. The energy constraints require that $c_{11} > c_{66} > 0$ (Dellinger 1991), so that negative c_{13} leads to negative ν_V. Physically it means that if you applied an axial stress along the TI symmetrical axis, the TI medium will shrink along the bedding. This is intuitively irrational for organic shales.

The real problem, I found later, is that I had made the same error Wang (2002) made: the velocity I measured was actually group velocity, not phase velocity. I made the correction, and all the negative c_{13} values become positive.

The lesson here is that unless the precise details of the experiment have been published — and you have read them — you should approach all data with caution. Learn about lab methods, ask questions of experimentalists and, if the data are important or look unusual, talk to the person that made the measurements.

References

Dellinger, J A (1991). *Anisotropic Seismic Wave Propagation*. PhD thesis, Stanford University.

Wang, Z (2002). Seismic anisotropy in sedimentary rocks, part 1: A single-plug laboratory method. *Geophysics*, **67** (5), 1415–1422. DOI: 10.1190/1.1512787.

Pore pressure and everything else

Fernando Ziegler

I want to highlight some of the tools rock physicists use to calculate the overburden and pore pressure, along with other useful things, such as fracture pressure, minimum horizontal stress, wellbore stability, seal capacity, and hydrocarbon column heights. While various disciplines come together to derive all the elements of this process, my goal is to describe only those elements in which rock physics is involved.

To start, the *overburden pressure*, or total vertical stress, is generated by the combined weight of the overlying rock matrix and the fluids filling its pore spaces. This is calculated from a composite bulk density log. Formation density measurements are sometimes incomplete for the overall borehole section, and in these cases rock physics-based empirical relations or effective field theory models are used to estimate density, such as from compressional wave velocities. We can also use data from nearby wells, calibrated to the specific lithology for the zone of interest.

Pore pressure can be calculated from direct or effective stress methods. Effective stress methods for pore pressure prediction are based upon Terzaghi's effective stress principle. In general, all methods are specific to particular areas and require further calibration to available measured pressure measurements and drilling events.

To properly model pore pressure, we also need to look at several of the mechanisms which generate these pressures. Shale mineralogy and thermal history, for example, play important roles as compaction and temperature change with burial. We also need rock physics models to account for other mechanisms, such as clay diagenesis, where a transformation of smectite to illite is observed at higher temperatures, or unloading through erosion which leads to a reduction in effective stress.

Fracture pressure, or more specifically the minimum fracture extension pressure, and *minimum horizontal stress*, or more specifically the fracture closure pressure, can be calculated using the previously derived overburden and pore pressure and by estimating the rock's mechanical properties, such as Poisson's ratio and internal friction angle. In addition, consideration must be taken in calculating magnitudes depending on the stress regime.

Wellbore stability analysis also uses some of the same estimated mechanical properties to determine shear failure stresses. Like pore pressure, wellbore stability also requires the calibration to available data. The petrophysical interpretation of image logs, where drilling induced breakouts may be present, is often useful, as are drilling events which help determine the magnitude and direction of the maximum horizontal stress. Then the pore pressure and wellbore stability analyses are coupled together and both are used to determine the minimum mud weight which will be required to prevent drilling problems along the proposed wellbore.

Finally, an understanding of capillary pressure and the physics of displacing a non-wetting fluid through the pores of a rock saturated with a wetting fluid is key for understanding *seal capacity* — the force required for hydrocarbons to flow through the seal — and calculating *hydrocarbon column height estimates*. This is done by examining mercury injection capillary pressure curves and looking for the characteristics of a seal, such as higher displacement pressures and higher wetting phase saturations. These results are then converted to apply to formation water and oil or gas equivalents. Thus the maximum hydrocarbon column height that the seal can hold is estimated. In the absence of mercury injection capillary pressure, other dynamic mechanisms are investigated to determine whether or not the reservoir pressure exceeds the fracture pressure of the seal and hydrocarbons have leaked off.

In general, rock physics isn't just about computing rock and fluid properties from well logs and seismic data. There are a lot of other applications, even more than those I have just described. The fun of it comes from the elements we borrow from other disciplines such as geology, geophysics, petrophysics, geomechanics, drilling, and completions. Or maybe, to put it another way, from the fact that rock physics is a key component in these disciplines.

References

Mavko, G, T Mukerji, and J Dvorkin (2009). *The Rock Physics Handbook: Tools For Seismic Analysis of Porous Media*. Cambridge University Press.

Terzaghi, K, and R B Peck (1948). *Soil Mechanics in Engineering Practice*. John Wiley & Sons, New York, NY.

Pressure signals are everywhere

Dan Ebrom

Pore pressure prediction is a key need in the oil industry. Accurate prediction helps avoid environmental contamination, improves the operational safety for the crew, and reduces the time needed to drill wells. A 3D pressure prediction can also help understand the likely migration paths for hydrocarbons in the subsurface, and thus evaluate charge risk for individual prospects.

Rock physics describes the relationships between a rock's velocity, density, and resistivity to its lithology, pore pressure, and pore fluid composition. While most hydrocarbon exploration applications of rock physics attempt to predict properties of reservoir bodies (clean sands, carbonates, etc.), pore pressure analysis is geared towards an understanding of the overburden, i.e. shales, shaly sands, and sandy shales.

For pore pressure, the fundamental input is velocity (generally P-wave velocities from 3D imaging). Velocities are more robust than seismic amplitudes, but also are lower in resolution than seismic events. Higher pore pressures are expressed through lower interval velocities.

Pressure prediction from seismic velocities allows the estimation of pore pressure at any point in the earth where a reliable velocity can be extracted. But lithology variations can cause perturbations to velocity (such as increasing sand content leading to higher velocity and erroneously lower predicted pressures).

We can better understand these perturbations through rock physics regressions such as the Eberhart-Phillips–Han–Zoback equations (EHZ):

$$V_\mathrm{P} = 5.77 - 6.94\varphi - 1.73\sqrt{C} + 0.446(P - e^{-16.7P}),$$

$$V_\mathrm{S} = 3.70 - 4.94\varphi - 1.57\sqrt{C} + 0.361(P - e^{-16.7P}).$$

EHZ uses simple algebra to predict the P-wave or S-wave velocity of a clastic sedimentary rock with a specified volume fraction of clay C, porosity φ, and a particular effective stress P in kilobars. The clever thing about the regression is that it can be used to predict the transform exponents (Eaton exponents) to map velocity measurement to pressure.

When I worked at BP, I became interested in the utility of S-wave velocities

While most hydrocarbon exploration applications of rock physics attempt to predict properties of reservoir bodies, pore pressure analysis is geared towards an understanding of the overburden.

for pore pressure estimation. My interest in S-waves stemmed from my PhD advisor, Bob Tatham, one of the pioneers in multicomponent seismology. He encouraged me to look for opportunities to tease extra information out of the generally neglected 'secondary' modes of wave propagation. P-wave reflectivity is king when it comes to mapping, but lithology information can be found in many other parts of the seismic wave field. EHZ predicts that S-wave velocities should be significantly more sensitive to pore pressure variations than P-wave velocities. This is expressed by a lower Eaton exponent for S-wave velocities (about 2) than for P-wave velocities (about 3). (Resistivity, which has the highest sensitivity to pressure for commonly measured properties, has an Eaton exponent of about 1.) S-wave velocities are now a common tool for detecting pressure in unconventional gas reservoirs because the measurements of S-wave velocities are unaffected while P-wave velocities are lowered in the presence of gas.

Also at BP, I worked with Martin Albertin and Philip Heppard to predict pore pressure beneath massive salt bodies, where the velocities from reflection seismic data were poorly known. Using 3D VSPs, championed by Brian Hornby, we captured reflected P to S mode conversions and calculated V_p/V_s ratios. The calculation method was proposed by Leon Thomsen and demonstrated by Mike Mueller. The simple rule is that V_p/V_s ratios decrease with depth, unless overpressure is present, so we had a tool to consistently predict subsalt overpressure — this information could be used to better drill the well as the salt–sediment interface was crossed.

The main message here is that opportunities to apply rock physics are everywhere, especially where no one has yet regarded a geological problem (like pressure prediction) as susceptible to rock physics analysis.

References

Eberhart-Phillips, D, D-H Han, and M D Zoback (1989). Empirical relationships among seismic velocity, effective pressure, porosity, and clay content in sandstone. *Geophysics* **54** (1), 82–89. DOI: 10.1190/1.1442580.

Ebrom, D (2003). *Pressure prediction from S-wave, C-wave, and P-wave velocities.* SEG Expanded Abstracts.

Viceer, S, M L Albertin, G Vinson, B Hornby, D Ebrom, P Heppard, C Jay and J-P Blangy (2006). *Improved drilling efficiency using a look-ahead VSP to predict pressure exiting salt: five Gulf of Mexico examples.* OTC Proceedings 18262.

Quantitative interpretation or quantitative everything?

Lee Hunt

Is it more important to be correct or right? One would think that they would amount to the same thing, but there are no perfect answers in our work. When we are making operational decisions, the debate is a question of usefulness, and the geoscientist must understand the broader business objectives within which any estimates are used. But embracing utility does not mean we should stop being scientists. Indeed usefulness calls us to apply the scientific method to a broader variety of questions.

We must apply quantitative methods to everything we do, not just to rock physics. Avseth et al's (2005) book *Quantitative Seismic Interpretation* is an impressive treatise on applying exploration rock physics in a quantitative way. But some people think that quantitative interpretation is just about rock physics. We must apply quantitative methods to *all* interpretive objectives. Everything we do should be quantitative.

A wrong but useful answer is better than nothing. We work almost entirely in an inductive, probabilistic, error-filled environment. We seldom have a perfect answer to give. We usually lack the data with which we may give such an answer. But we must try for an approximation that is *useful enough*: because useful enough is still useful. Geomechanics is fraught with this dissonance. For instance, some practitioners see the issues of static versus dynamic rock properties and argue that we cannot make geomechanical estimates using log data. There is no argument that it is better to make both static and dynamic rock measurements. But we must not fail to do our best with what data we have.

To find the right questions we have to be prepared to frame our problems in a practical manner. Problem solvers in an industrial setting must ask what the real questions are. *The end questions*. Are we trying to estimate rock physical properties because that is where economic value directly comes from, or are we trying to infer some other things from the rock properties? Our objectives are ultimately economic, which means we are considering production, costs, and risks. This often leads us to consider Darcy's law, which states that flow capacity is a product of permeability, pressure draw-down, mobility, and an area term. Rock physics is rarely the primary objective of the exercise, but it can often be critical in inferring permeability, or the area term via geomechanics (frackability), or even pressure by the rocks' response to pressure changes. This places rock physics on the one hand as having crucial importance, while on the other hand being unimportant in itself.

Other times, we fail completely to do what we set out to do in the first place, but still achieve our final objective of adding value. Hunt et al (2012) could do nothing to predict the Saskatchewan Manitoba Three Forks reservoir quality. The reservoir was too thin, and sat under a highly reflective shale. Instead we added value by reprocessing the data to a point where accurate depth mapping was possible. The depth mapping was used to steer wells more accurately. Greater accuracy in steering reduces the risk of catastrophic operational failure, reduces operational costs through reduced sidetracking, and increases production through greater access to the reservoir area.

Hunt et al (2014) demonstrated a complete failure of AVO to yield accurate results on the Wilrich tight gas sands in Alberta. Despite that failure, we did quantitatively demonstrate that steering accurately into the most permeable reservoir element was strongly related to well productivity. This oriented the team towards the value of accurately steering the horizontal wells in that area. In order to prove this, the team was required to work with as much geological and engineering data as geophysical data. Quantitative interpretation required a quantitative look at everything. In order to best add value in these two examples, we had to disengage ourselves from the mindset of the geophysicist acting within a narrowly defined role, and consider the larger role of problem solver. It was crucial that we determined what data could actually be used and did not confine ourselves to our previously held paradigms.

Don't let quantitative interpretation mean a narrow, confining, thing: make it quantitative *everything*. We add value best when we consider the larger, more general aims of our work, and remember that as generalists we have to ask the broad questions as well as the deep ones.

References

Avseth, P, G Mavko, and T Mukerji (2005). *Quantitative Seismic Interpretation: Applying Rock Physics Tools to Reduce Interpretation Risk*. Cambridge University Press. 409 pages.

Hunt, L, R Reynolds, S Hadley, and J Downton (2012). Quantitative Interpretation part II: case studies. *CSEG Recorder* **37** (2), 44–54.

Hunt, L, S Hadley, S Reynolds, R Gilbert, J Rule, and M Kinzikeev (2014). Precise 3D seismic steering and production rates in the Wilrich tight gas sands of West Central Alberta. *Interpretation* **2** (2), 1–18. DOI: 10.1190/INT-2013-0086.1.

Right, wrong, and useful rock physics models

Ran Bachrach

> *Essentially, all models are wrong, but some are useful.*
> **George Box (1987)**

This quote has been used in the rock physics community for some time to justify the use of a model when challenged about its assumptions. The reader should note that Box was writing about statistical empirical model building.

Rock physics models can be categorized into three main categories:

1. Theoretical models
2. Phenomenological models
3. Empirical models

A theoretical model must be satisfied when the assumptions are met. Examples of such models are the Hashin–Shtrikman bounds and Gassmann's equation. These models must be honored such that if you see data that violates them you know that your current understanding of the rock in terms of its mineralogy, porosity, fluid moduli, and grain compressibilities or pore pressure cannot be right. For example, if we observe that a rock mixture, say a binary brine–quartz mix violates the isotropic Hashin–Shtrikman bounds we know that either the constituents are not brine and sand, or that the mixture is anisotropic. To some extent, the theoretical models will show us if our state of knowledge of the rock is correct. These models will never be wrong if their assumptions are met, and the assumptions made about these models are often met in real rocks. For these models, the uncertainty is not in the model itself but only in the assumptions and input variables.

Phenomenological models try to model the rock using theoretical results based on simplified assumptions which are almost never met in the real rocks, but can be used in a predictive manner. Examples of such models are the well-known Kozeny–Carman relationships between porosity and permeability, which are exact for a simplified network of pipes but are extended to describe permeability of real rocks based on a dimensional similarity argument. Another example is the family of inclusion models based on Eshelby's single inclusion result, which is exact for ellipsoidal inclusions, but is used with different homogenization schemes to describe the behaviour of porous media. These models are based on theoretical results but are 'wrong' in the sense that the geometries they describe are practically never met in real rocks. Therefore, I tend to categorize this subset of models as phenomenological models which show that, for example, idealized objects such as pores modelled as ellipsoidal inclusions, have a certain behaviour that can be expected in real rocks. Phenomenological models are useful because they enable us to understand rocks which cannot be explored otherwise. The uncertainty associated with this type of model is not only related to the rock constituents but also to the micro-mechanical/geometrical parameters used.

An empirical model is a functional form whose parameters are usually derived from regression analysis of data points. Note that the functional form here is often a simple mathematical object (polynomials or power laws, e.g. Han's regressions, Gardner's relations, the mud-rock line, etc.). Because these types of models are empirical, different statistical methods can be used to quantify their performance and prediction error, and the model can be used to characterize the population it was sampled from assuming the sampling is a representative subset of the formation of interest. Calibration of the functional model is always data specific, and in principal there are no direct ways to ensure the set of regression coefficients or the functional form will be valid on other data sets. Clearly, model uncertainty is well defined on the sampled population, but poorly extrapolated beyond the bounds of the measurements or to other datasets.

Not all rock models carry the same validity or uncertainty. Next time you apply one, ask yourself 'What kind of model is this?'.

References

Box, G, and N Draper (1987). *Empirical Model Building and Response Surfaces*, John Wiley & Sons, New York, NY.

Shale rock physics

Dave Dewhurst

The majority of rocks in most sedimentary basins are shales of one form or another. Their properties have historically been ignored as they were not considered as economically valuable. If anything, overburden shales have long been associated with problematic wellbore instability, high pore pressures, and anisotropic complexity. About a decade ago, the gas shale bandwagon pulled into town and, with the economic potential now raised, there has been a scramble to establish credentials in shale rock property testing. However, there are critical issues for shale rock physics testing which are not widely understood, and which are often ignored for the sake of simplicity or saving time.

At first sight, overburden shales and gas shales appear to be two different things. The former are usually clay-rich and often organic poor with their rock properties governed by the clay matrix. Gas shales on the other hand tend to be dominated by either carbonates (calcite, dolomite) or rigid silicates (quartz, feldspar) and contain significant volumes of organic matter, although many do contain clay minerals (10–40 percent, usually illite and chlorite). It is the clay-bearing nature of these materials together with nanometre-scale pore throat sizes that cause problems for geomechanical and rock physics testing for shales.

Probably the most critical issue for the determination of experimental rock physics properties of shales is understanding the saturation state of the material before testing. Preservation of shales from the moment of recovery is critical, as the desiccation of cores will radically alter shale properties. Figure 1a shows the impact of decreasing water saturation from 100 to 40 percent on rock physics response of a silty shale (~30% clay), with dynamic bulk modulus decreasing significantly while dynamic Young's and shear moduli rise. The shear modulus increase is about 20 percent and much more significant than the usual very slight changes noted between saturated and dry coarser-grained rocks. The figure on the right shows the increase in strength and elastic modulus of a clay-rich shale (~60% clay) due to decreasing water saturation. The nanometre scale pore throat sizes in these materials mean we are in a world dominated by capillary and osmotic forces that is not controlled by simple Terzaghi effective stress laws. Capillary and osmotic suctions set up by decreasing water saturations in such materials can cause increases in effective stress of tens of MPa or

Do not apply laboratory measurements on dried-out shales
to downhole situations as the material is
no longer in a state even close to what it was in situ.

more (Laloui 2013), and these can account for the increases in shear modulus, static Young's modulus, and strength observed in the figure. The lesson: do not apply laboratory measurements on dried-out shales to downhole situations as the material is no longer in a state even close to what it was in situ.

A second significant issue for shale rock physics at both the laboratory and field scale is anisotropy, where velocities and elastic moduli vary with direction of wave propagation through the rock. This is caused by multiple sources from the flat, platy or needle shape of the clay particles themselves, preferential alignment of those particles due to compaction and/or diagenesis, to the presence of laminations and the shape and alignment of organic matter. Intrinsically, both overburden shales and gas shales are considered to be transversely isotropic, in that velocity parallel to bedding is higher than the velocity normal to bedding but equal in all directions. Natural fractures may alter this symmetry, however, especially in gas shales. In the laboratory, anisotropy is best measured using single core plug techniques with ultrasonic transducers oriented both parallel and normal to bedding plus transducers at off axis angles. Multiple off axis transducers are recommended to better constrain the full elastic tensor and the tricky-to-measure (in the lab) Thomsen's δ parameter.

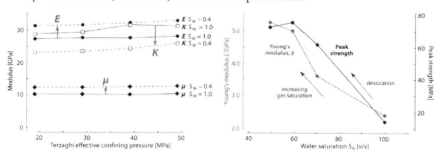

The figures show the change in dynamic elastic moduli (left) plus strength and static Young's modulus (right) as water saturation decreases in clay-bearing shales.

References

Laloui, L (2013). *Mechanics of Unsaturated Geomaterials.* Wiley, 382 pp.

Thomsen, L (1986). Weak elastic anisotropy, *Geophysics* **51**, 1954–1966. DOI: 10.1190/1.1442051.

Some do's and don'ts

Rob Simm

A combination of seismic interpretation and quantitative analysis skills is a powerful combination in oil and gas exploration today. A key obstacle for many geophysicists, however, is the time and effort required to establish a rock physics database relevant for the play at hand. As an aid to those willing to venture forth into the world of log-based rock physics, here are a few key do's and don'ts for conventional reservoirs.

Log conditioning

- Check for log data integrity. Check the caliper log for hole variations, then look for sympathetic variations in density and sonic. Make crossplots and compare results to 'typical' rock physics trends/relations.

- Apply realistic edits. Don't edit using straight lines, and beware of giving too much significance to models generated from logs where excessive editing has been done to the data.

- Model the water first. Leave the hydrocarbon wells alone until you have a good idea of the appropriate rock physics relations. Errors in fluid assumptions are more likely to be relatively insignificant in water.

- Calibrate V_S prediction functions to measured data. V_S prediction is effectively encoding AVO differences into the model. Most linear or quasi-linear V_P versus V_S functions are relevant for the wet case only; do not apply them to logs with hydrocarbons. For hydrocarbon cases it is best to apply a calibrated dry rock model and substitute brine before estimating V_S. An alternative is to use specific theoretical models (e.g. critical porosity model) to describe the 'stiffness' characteristics of the reservoir.

- Invasion is not always straightforward to address. Invariably the density log will be affected, particularly if the density contrast of the filtrate and virgin fluid is large. The sonic log may or may not show invasion effects. Fluid substitution can be used for invasion correction but, in order to constrain the solution, log data of high quality is required (including V_P, V_S, density, and a good idea of the porosity). If possible, compare the dry rock characteristics of hydrocarbon and water zones (but be aware that diagenetic differences might exist between the two zones).

- Evaluate fluid substitution effects — don't just push the button. Check the dry rock estimates and use templates to validate. After fluid substitution, if the change in V_p in shaley sands is greater than in the clean sands there is likely to be a problem. Check that the Poisson's ratio log looks reasonable, especially when substituting to gas.

- Consider the possible effects of anisotropy. For example, shales in deviated wells will have higher velocities than in vertical wells. If possible make simple models to evaluate the nature and magnitude of anisotropy on AVO responses.

Rock characterization and modelling

- Determine separate seismic lithofacies on the basis of acoustic crossplots (e.g. acoustic impedance versus Poisson's ratio) and take into account factors such as mineralogy and stiffness. Do not, for example, average clean sand and shaley sand data.

- If possible, derive probability density functions from the data, but remember that in the first instance these are related to the log scale. Any models generated using these distributions should be appropriately upscaled.

- Apply fluid substitution before upscaling.

- Beware of averaging across thin beds. The results of simple, single-interface AVO models in wells with thin beds may not be representative.

- Be wary of predicting outside the range of the data. Do not place too much faith in the rock physics model, particularly if it is based on a small amount of measured data.

- Uncertainty cannot be adequately gauged by looking at the errors in a small number of wells. It is quite possible that there is bias in the well points, e.g. wells may only have been drilled in the zones with bright amplitudes.

- Uncertainty arises from geological variance, calibration uncertainties, and seismic errors. Geological variance is a key issue requiring a rock model which describes the lithofacies statistics (i.e. V_p, V_s, and density distributions and covariance relations), as well as layer thicknesses and interface combinations.

- Think carefully when invoking a seismic attribute/reservoir property relation. Use synthetic data, modelled from wells, to first derive the relationship (see *Attributes on the rocks* on page 20–21). When plotting mapped attribute values at well locations against synthetic values, remember that the nearest sample to the well location may not always be the most representative, so plotting a range of mapped values (e.g. nearest nine grid points) may help in determining an appropriate scale factor.

Some thoughts on the basics
Tad Smith

On the airplane home from a recent SEG convention I took the opportunity to make some notes for this essay rather than snoozing and listening to some good music (and believe me, I could have used some shut-eye).

The convention was, as always, a great time to learn a few new things, catch up with old friends, and do some networking. The Rock Physics Reception in particular is a great opportunity for meeting fellow rock physics enthusiasts. I'm always struck by how far rock physics has come in the 15 or so years that I have been engaged in the practical application of the discipline. The explosion of new ideas and new models is truly impressive. However, when I compare some of the higher-end theoretical work to what we do at the 'rock face', I come to the conclusion that the workhorses of the discipline really haven't changed much in many years, and I don't expect them to change much in the future.

Some things are timeless and are part of the foundation of what we do. The need to have, for example:

- properly conditioned log data,
- a calibrated and meaningful petrophysical model,
- proper shear velocity edits and/or estimates,
- and properly constructed fluid substitution models that don't violate any of our normal constraints (e.g. negative dry frame properties).

I've written about these things elsewhere, and have given numerous talks on the need to properly handle the basics. Personal observation and experience remind me this is a message worth repeating many times. Unfortunately, properly executing some of the more mundane parts of the seismic petrophysics workflow can be a painful process, and often takes more time than we can reasonably spend on a project. Unfortunately, this leads us all (myself included), to sometimes take compromising shortcuts. As busy as I am these days, I must frequently remind myself that getting the fundamentals right is well worth the investment in time.

So what's new that will 'move the needle'? Although I don't think the fundamentals of our discipline will ever change, the rapid growth of organic-rich

...Getting the fundamentals right is well worth the investment in time.

shale plays has forced me to develop an understanding of anisotropy that I was heretofore happy to ignore. Leon Thomsen has been beating the anisotropy drum for many years, but most of us have been toiling away under the assumption of isotropy. This has worked reasonably well for many years and for many projects in conventional rocks. However, the more I have studied organic-rich shale, the more I have become convinced that we can't begin to understand them without considering the effects of anisotropy. You certainly can't reliably estimate a shear log in these rocks using conventional techniques. Be exceedingly cautious if you are using Greenberg and Castagna to estimate a shear log in an organic-rich mudstone!

It's my firm belief that the need to do the fundamentals of seismic petrophysics properly will never go away. Nor can it be automated or farmed out to a technologist. However, while you're working hard to get the fundamentals right, don't forget to take the time to keep up with some of the newer — but complicating — aspects of rock physics.

References

Greenberg, ML, and JP Castagna (1992). Shear-wave velocity estimation in porous rocks: Theoretical formulation, preliminary verification and applications. *Geophysical Prospecting* **40**, 195–209. DOI: 10.1111/j.1365-2478.1992.tb00371.x

Thomsen, L (1986). Weak elastic anisotropy, *Geophysics* **51**, 1954–1966. DOI: 10.1190/1.1442051.

Sonic logs are not true

Xinding Fang & Michael Fehler

Monopole and cross-dipole sonic logs are widely used for determining formation P-wave velocity, and S-wave velocity and anisotropy respectively (see *How to catch a shear wave* on page 50–51). Because borehole sonic logs are direct measurements of the subsurface, they are taken as true measurements for constraining reservoir velocity models inverted from surface seismic data. However, drilling a borehole strongly alters the stress distribution in its vicinity. Drilling-induced stress changes lead to additional stress-induced elasticity anisotropy in the formation that affects the velocities measured by sonic logs. Moreover, the redistribution of stress around a borehole may change the shape of the borehole because of mechanical damage. Clearly, understanding the effect of stress on sonic measurements is key to interpreting sonic log data.

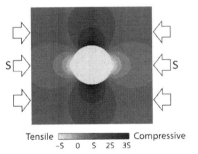

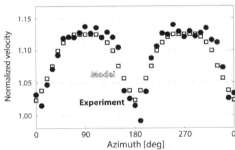

The figure on the left shows the variation of the sum of principal stresses when a uniaxial stress S is applied normal to the borehole axis. Dark and light colours respectively indicate regions of stress concentration and relief. In the compressive regions, the stress can be up to three times the loading strength and can substantially increase the rock stiffness, inducing borehole breakouts if it exceeds the rock strength (Zoback et al. 1985). Low velocity zones appear in the tensile regions because of stress-induced crack opening. Since most sedimentary rocks exhibit stress-dependent elastic responses, the variation of stress due to borehole alteration results in both azimuthal and radial variations in formation elastic properties.

Winkler (1996) experimentally studied the influence of stress compression on borehole P-wave velocity (V_p) measurements. In his experiment, uniaxial stress was used to effectively represent the differential stress normal to borehole axis in the earth. The figure on the right shows the azimuthal variation of V_p (normalized by the velocity before stress loading) in a Berea sandstone sample with a borehole at the centre and 10 MPa uniaxial stress. Also shown is the modelling result for the experiment (Fang et al. 2014). The P-wave velocity at a given azimuth is measured using a directional transducer and two receivers to excite and record the P-wave. Both experimental and numerical results show that V_p has maxima normal to the stress direction and minima parallel to it. The strong variation of V_p in the figure indicates that the velocity determined from sonic measurements might deviate from that of the virgin formation due to borehole stress alteration. Bear in mind that for a typical sonic tool, you may measure an average of these azimuths.

In summary, concentration of stress around a borehole affects sonic logging measurements by changing the near wellbore formation elastic properties that could bias velocity measurements.

References

Fang, X, M Fehler, and A Cheng (2014). Simulation of the effect of stress-induced anisotropy on borehole compressional wave propagation. *Geophysics* **79** (4), D205–D216. DOI: 10.1190/geo2013-0186.1.

Winkler, K (1996). Azimuthal velocity variations caused by borehole stress concentrations. *Journal of Geophysical Research* **101** (B4), 8615-8621. DOI: 10.1029/96JB00093.

Zoback, M, D Moos, L Mastin, and R Anderson (1985). Well bore breakouts and in situ stress. *Journal of Geophysical Research* **90** (B7), 5523–5530. DOI: 10.1029/JB090iB07p05523.

Take the time to explore relationships

Wes Hamlyn

One of the many things I love about rock physics is how several relatively straightforward concepts can be combined for more complex and powerful uses. One such example that comes up frequently, for me at least, is the concept of pore space stiffness. Let's take a look at three fairly simple ideas related to pore space stiffness and how they can be used together to gain geological and geophysical insights.

1. Mavko and Mukerji (1995) presented the compressibility of a dry rock as the combination of mineral and pore space compressibilities. Written in terms of bulk modulus, this relation is expressed as

$$\frac{1}{K_{dry}} = \frac{1}{K_0} + \frac{1}{K_\varphi},$$

   where K_{dry} is the bulk modulus of the dry rock, K_0 is the bulk modulus of the mineral grains, and K_φ is the bulk modulus of the pore spaces.

2. Filling the pore space of a dry rock with fluid has the effect of increasing the pore space stiffness by an amount, F, where F is a function of the fluid and mineral bulk moduli. This is expressed as

$$\frac{1}{K_{sat}} = \frac{1}{K_0} + \frac{1}{K_\varphi + F}.$$

3. By normalizing the rock compressibility by the mineral compressibility $(1/K_0)$ we obtain the expression

$$\frac{1}{K/K_0} = 1 + \frac{1}{k},$$

   where K may be K_{dry} or K_{sat} and $k = K_\varphi/K_0$. From this normalized bulk modulus we can gain insight into whether a rock is stiff or compliant.

So, let's put these concepts to use. The figure opposite shows data from two sand intervals, one shallow (circles) and one deep (squares). Bulk moduli have been calculated from V_P, V_S, and density logs for in situ (solid points) and dry rock (open points) conditions using Gassmann substitution. All data have been normalized by the mineral compressibility, $1/K_0$.

Overlain on this plot are contours of constant pore space stiffness, k. Small values of k correspond to softer rocks while larger values of k correspond to stiffer rocks.

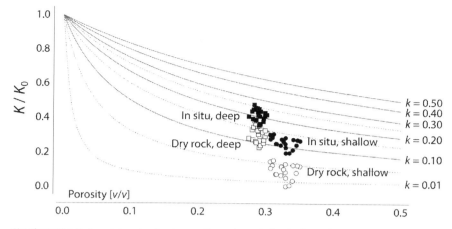

Plot of normalized bulk modulus values from two sand intervals, one shallow (circles) and one deep (squares). Bulk moduli have been calculated from V_p, V_s, and density logs for in situ (solid points) and dry rock (open points) conditions using Gassmann substitution.

Comparing the shallow and deep sands at dry rock conditions (open points), we note that the two sands are centered on different contours. The deep sands plot along a larger stiffness contour (~0.13) than the shallow sands (~0.03). From this we infer that the deep sands are stiffer and less compressible than the shallow sands.

We noted previously in point (2) that adding pore fluid to a dry rock has the effect of increasing the pore space stiffness. So adding fluid to the dry rock will cause points to shift vertically an appropriate number of pore space stiffness contours.

In our example, Gassmann substitution was performed using the same pore fluid for both sands. This means that the number of pore-space stiffness contours jumped will be the same for both sands. But, because pore-space stiffness contours are not equally spaced, the magnitude of the change in bulk modulus will not be the same for both sands. Naturally the softer sands will show a larger fluid effect. This is supported by the large spacing between pore-space stiffness contours in the figure.

There are many similar simple relationships in rock physics that, when combined, allow us to explore interesting scenarios, and tune our intuition for the next well.

References

Mavko, G and T Mukerji (1995). Seismic pore space compressibility and Gassmann's relation *Geophysics*, **60**, 1743–1749. DOI: 10.1190/1.1443907.

Texture matters

Colin Sayers

Sedimentary rocks are composed of a rock framework consisting of various grains and porosity that may contain various fluids such as free and clay-bound water and hydrocarbons. Rock physicists require estimates of the composition of sedimentary rocks as input to effective-medium theories such as the self-consistent scheme and the differential scheme, or effective-field theories such as the Mori–Tanaka scheme. These theories are used to compute seismic wave velocities, rock properties, such as bulk and shear moduli, and to investigate the sensitivity of elastic wave velocities to fluid saturation, pore pressure, stress, temperature, and other variables.

In many cases, estimates of the composition of the rock are provided to rock physicists by petrophysicists, who estimate the composition of the rock by analyzing log measurements such as gamma-ray, density, neutron, resistivity, and sonic. Often, this is done using a multi-tool inversion to solve for the volume fractions of the various components from a number of different well logs. However, this is not enough.

A full description of rock microstructure requires more than just a knowledge of the volume fractions of its components. Also important is the rock fabric, or texture, which involves the shapes of the individual grains as well as their spatial arrangement, interconnectivity, etc. Unless the arrangement of the components is very simple, as, for example, a 1D layered medium, it is extremely difficult to calculate the properties of the rock exactly. For macroscopically isotropic rocks, upper and lower bounds to effective properties were provided by Hashin and Shtrikman (1963) for the case in which only the volume fractions of the components are specified. But closer bounds can be obtained given information on the spatial arrangement of the components (Torquato 2002). This simple example shows the upper bound for the bulk and shear moduli of an elastic medium containing a spherical distribution of randomly oriented dry circular cracks, i.e. a spatial distribution in which each crack is surrounded by a spherical region into which no other crack is allowed to penetrate. The curves are plotted as a function of crack density which, for penny-shaped cracks of radius a, is defined to be na^3 where $n = N/V$ is the number N of cracks in a representative volume V.

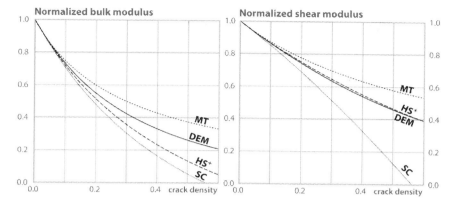

A comparison of upper bound of Ponte Castañeda and Willis (1995) (HS⁺) with the predictions of the self-consistent (SC), differential (DEM), and Mori–Tanaka (MT) schemes for the normalized bulk and shear moduli of an elastic medium with Poisson's ratio v = 0.25 containing a spherical distribution of randomly oriented dry circular cracks. Figure redrawn from Ponte Castañeda and Willis, 1995.

This was calculated using the method of Ponte Castañeda and Willis (1995), based on the Hashin–Shtrikman variational principle, that allows the effect of particle shape and distribution to be included, thus resulting in tighter bounds. Also shown are the predictions of the self-consistent, differential, and Mori–Tanaka schemes. Since the Hashin–Shtrikman estimate shown by the dashed curve is an upper bound, it follows that the differential and Mori–Tanaka schemes do not give a good description of the elastic stiffnesses for this particular microstructure, while the predictions of the self-consistent scheme are consistent with the bounds. Thus the spatial distribution of the various components of the rock have an important effect on the properties of the rock, and the various rock physics models contain assumptions, either explicit or implicit, about the microstructure of the rock.

References

Hashin, Z and S Shtrikman (1963). A variational approach to the theory of the elastic behaviour of multiphase materials. *Journal of the Mechanics and Physics of Solids* **11** (2), 127–140. DOI: 10.1016/0022-5096(63)90060-7.

Ponte Castañeda, P and J Willis (1995). The effect of spatial distribution on the effective behaviour of composite materials and cracked media. *Journal of the Mechanics and Physics of Solids* **43**, 1919–1951. DOI: 10.1016/0022-5096(95)00058-Q.

Torquato, S (2002). *Random Heterogeneous Materials Microstructure and Macroscopic Properties*. New York: Springer.

The astonishing case of non-linear elasticity

Paul Johnson

Imagine the following. Tap a bell and it rings at one or more resonance frequencies. Tap it harder, and it simply rings louder. Now, tap a bell that contains a crack. It too rings normally. But as you tap it progressively harder, the ringing frequency progressively decreases. The crack causes the elastic properties that control the ringing frequency to change with amplitude.

The same is true of a sample of rock. As you tap it progressively harder the ringing frequency progressively decreases. This is because rocks are naturally damaged at many length scales. If you could hear the ringing of a sand pile when it is 'tapped' by an acoustical wave emitted from a loudspeaker, you would find the same behaviour. As the amplitude increases, the ringing frequency decreases. (Because the sand pile attenuates sound so well, and because the tapping frequency may be below the audio range, you would not hear this, but a sensitive detector would.) The grain bonds respond in the same manner that a crack does. In short, the speed of sound through a material is a measure of the material's modulus (the relation between applied stress and detected deformation) and density. The speed of sound as a function of wave amplitude is a measure of the material's non-linear elasticity, primarily related to intrinsic damage and bond integrity.

Exotic elastic behaviour is found in cracked metals, rocks, sand piles, and more. However, if you conducted the same experiment with a rod of steel, you would find that the ringing frequency is independent of the amplitude. Undamaged steel is linearly elastic. Rocks, cracked steel, and sand piles are non-linearly elastic. What does this mean? It means that the modulus and the material's acoustic wave velocity are amplitude dependent, in contrast to linearly elastic materials. They are non-linearly elastic because the acoustic wave speed is dependent on the amplitude.

As it turns out, all solids, liquids, and gases exhibit very weak non-linear elastic behaviour — so weak that it is difficult to measure. This weak non-linear elasticity is due to atomic vibrations that are perturbed by a sound wave and induce distortion of the sound wave resulting in overtones, but not sound speed change. In contrast, in rock, damaged steel, and a sand pile it is the damage — cracks, fissures, and grain contacts — that cause their exotic non-linear elastic behaviour. We call it exotic because the effect can be strong. For instance, in a

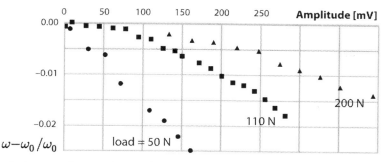

Normalized resonance frequency change vs detected amplitude for three applied loads in a glass bead pack. See Johnson and Jia (2009), Nature **437**, p 871.

sand pile the sound speed can decrease by more than five percent in compressional wave velocity at modest sound amplitudes, and we observe ~30 percent changes in shear wave speeds. That is remarkable! We also call it exotic because if one unravels the behaviour of a pressure wave versus its detected deformation, it contains many wave harmonics or overtones, and exhibits hysteresis in applied pressure versus detected deformation. Undamaged steel, single crystals, etc., do not behave this way. Moreover, when non-linearly elastic materials are subjected to static stressing, they exhibit strong changes in their wave velocity. Further still, if the static stressing is progressively applied and then progressively removed the change in sound speed that occurred as pressure is applied, is different than it is when the stress is removed. This is called stress-strain hysteresis.

Because damaged and granular materials behave in characteristic and easy to measure ways, one can apply acoustical methods to distinguish between undamaged and damaged materials in an extremely sensitive manner. This opens up a broad area of applications to industrial problems where one would like to know if a material is damaged. There are applications to the earth, too. For instance during earthquakes, soil layers at the earth's surface experience strong shaking and as a result go non-linear. The soil sound speed and sound dissipation can de reduced by more than half! This behaviour has strong implications for characterizing building sites and building structures that may experience a large earthquake. Furthermore, fault gouge material located between the fault blocks — the material that is created through crushing and grinding during repeated faulting episodes — can also be disproportionately affected by seismic waves. As a result faults can be triggered by seismic waves at distances of thousands of kilometres. We find that one of the explanations for triggering is the non-linear elastic behaviour of the gouge material due to the seismic waves. These cause the gouge material to weaken and fail.

In summary, non-linear elastic behaviour is very common and occurs at scales from millimetres to hundreds of kilometres. By characterizing these behaviours we can learn a lot about material properties that cannot be otherwise gleaned.

The dynamic to static correction

Dick Plumb

Geomechanical models help engineers minimize the economic impact of rock deformation when exploring for and producing oil and gas. Models based on dynamic elastic moduli show trends in mechanical properties, but do not accurately predict rock deformation. Because rocks rarely behave as linear elastic solids, a dynamic to static correction is required for accurate predictions. The correction pertains to all elastic moduli, including Young's modulus (E_d — the $_d$ denotes dynamic) and Poisson's ratio (ν_d), however this discussion focuses on Young's modulus as it has greatest bearing on geomechanical predictions.

In the context of linear elastic isotropic rock, the dynamic Young's modulus E_d can be computed from compressional wave velocity V_P, shear wave velocity V_S and bulk density ρ_B:

$$E_d = \rho_b V_S^2 \left(\frac{3V_P^2 - 4V_S^2}{V_P^2 - V_S^2} \right)$$

Static moduli are computed from laboratory tests on cylindrical rock samples deformed under drained conditions. The figure below shows stress–strain curves recorded in a drained triaxial test designed to measure static and dynamic moduli. On the right are three ways to measure Young's modulus: static loading (1), small unload–reload cycles (2), and from elastic wave velocity measurements (3):

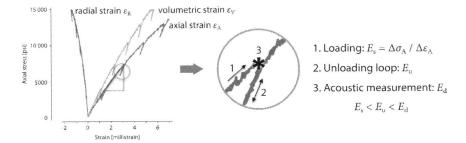

We see that the value of Young's modulus depends on how it is measured. Differences between E_s, E_u, and E_d represent the rock's response to different loading conditions.

E_s and E_u are determined from the changes in stress and strain, but E_d is determined from high-frequency elastic waves (~500 kHz) that exert low levels of transient or dynamic stress (<10^{-5} MPa). So long as dynamic strains are <10^{-6}, rock behaves like a linear poroelastic solid (Winker et al. 1979). Velocity dispersion associated with the Biot squirt-flow mechanism accounts for a few percent increase of E_d, over E_s. However, the effect is negligible in dry or gas saturated rocks or in saturated rock with very low porosity (Dvorkin et al. 1994).

E_u is determined using lower frequency (0.01 Hz) and greater deviatoric loads (~10 MPa). Under such quasi-static conditions, rock behaves as a linear viscoelastic material. The viscoelastic response is attributed to time dependent deformation of compliant rock elements such as grain contacts, microcracks, clay minerals, or lenses of kerogen (Miller et al. 2013).

E_s is measured under the greatest deviatoric stress (~50 MPa). Greater applied stress leads to plastic deformation at grain contacts or along surfaces of microcracks, or plastic deformation of clay minerals or kerogen.

To assess the significance of the dynamic-to-static correction know that:

1. Dynamic moduli are systematically greater than static moduli.
2. E_d can be a factor of ten greater than E_s.
3. The difference between E_d and E_s decreases as porosity decreases, as the stress used to determine the moduli decreases, and as the mean effective confining stress increases.

If accurate geomechanical predictions are required, dynamic moduli should be transformed to static moduli. Laboratory measurements should be made to quantify the magnitude of the dynamic to static correction and to determine the compositional and textural elements of the rock responsible for the differences. For geomechanical applications where rock strains are <10^{-6}, a static-dynamic correction is not generally required. If strains are >10^{-6} and a dynamic to static correction is not applied, geomechanical computations will contain systematic errors which must be acknowledged when applying the model.

References

Dvorkin, J, R Nolen-Hoeksema, and A Nur (1994). The squirt-flow mechanism: Macroscopic description. *Geophysics* **59** (3), 428–438. DOI: 10.1190/1.1443605.

Winkler, K, A Nur, and M Gladwin (1979). Friction and seismic attenuation in rocks. *Nature* **277**, 528–532. DOI: 10.1038/277528a0.

Miller, D, R Plumb, R, and G Boitnott (2013). Compressive strength and elastic properties of a transversely isotropic calcareous mudstone. *Geophysical Prospecting* **61**, 315–328. DOI: 10.1111/1365-2478.12031.

The Hilterman approximation

Mark Sams

The Zoeppritz equations are often regarded as difficult, at least in the sense that it is hard to decipher what they mean. But they are simple in the sense of the terms involved. They indicate that the reflection coefficients at an interface are determined by the elastic properties (P-wave velocity, S-wave velocity, and density) of the materials either side of the interface and the angle of incidence of the incoming wave. Beyond that it is not simple to relate the reflection coefficients directly with any aspect of the properties or changes at the interface.

A number of approximations have been made to the Zoeppritz equations in order to make them easier to use and understand. These approximations usually concern the reflection of a P-wave to a P-wave and assume that the contrasts in elastic properties across the interface are small and that the angle of incidence is not too high. The simplifications do not necessarily enhance the interpretability of the equation in terms of identifying links between reflection coefficients and elastic properties, but they do have some features in common that are useful to appreciate.

Wherever the elastic properties occur in the equations they take the form of either the contrast of a property divided by the average property across the interface or the ratio of the average S-velocity to the average P-velocity. The other elements to the equations are functions of the angle of incidence. This indicates that the reflectivity is not just related to the contrast in elastic properties but is also dependent on the average or background elastic properties. So for the same elastic contrast, the higher the average properties, the smaller the reflection coefficient. This suggests that reflection coefficients should in general get smaller with increasing depth as the average impedance increases. The other feature in common is that when the angle of incidence is zero, the reflection coefficient is equal to one half of the contrast in acoustic impedance divided by the average acoustic impedance. That is, the zero offset reflection is dependent on acoustic impedance alone.

There is one approximation that stands out from the rest in terms of providing an easy-to-interpret formulation — the Hilterman approximation:

$$R(\theta) \approx \frac{\Delta I}{I} \cos^2\theta + \frac{\Delta v}{(1-v)^2} \sin^2\theta$$

Reflection coefficients should in general get smaller with increasing depth as the average impedance increases.

The equation, proposed by Verm and Hilterman (1995), contains terms in acoustic impedance I and Poisson's ratio v. In each case, the numerator contains the difference, and the denominator the average, across the interface. If the angle of incidence θ is set to zero, then the reflection coefficient R is a function of acoustic impedance alone. On the other hand, if the angle of incidence is set to 90° the reflection coefficient is a function of Poisson's ratio only.

Of course, setting the angle to 90° is meaningless in the real world, but it does allow us to interpret the character of AVO behaviour. For example, consider a shallow gas sand overlain by shale. Based on our general expectations of elastic properties for such rocks we would expect the gas sand to have a lower acoustic impedance than the shale and a much lower Poisson's ratio. This means that the zero offset reflection coefficient will be reasonably negative (note the background AI in the denominator is relatively low) and the 90° 'Hilterman' reflection coefficient will be strongly negative. Therefore we expect that the AVO response will be negative, getting more negative with increasing angle: the classic class 3 AVO for a soft gas sand. In a class 2p anomaly the zero offset reflection coefficient is positive and becomes negative at some intermediate angle. The equation implies that the layer below the interface has a moderately higher acoustic impedance but a lower Poisson's ratio, for example a hard gas sand.

The Hilterman approximation to the Zoeppritz equations is therefore a most useful formulation. It allows one to interpret and predict AVO behaviour in terms of elastic properties and is consequently a key to interpreting seismic amplitude data.

References

Verm, R and F Hilterman (1995). Lithology color-coded seismic sections: The calibration of AVO crossplotting to rock properties. *The Leading Edge* **14** (8), 847–853. DOI: 10.1190/1.1437170.

The message is the medium

Rob Lander

It amazes me that the micron-scale topology of rocks has a profound influence on acoustic properties at the seismic scale. Most rock physics studies, however, focus on the 'physics' and are hindered by inaccurate depictions of the 'rock'.

New models where the 'rock' gains the same elegance as the 'physics' of existing models would be vastly more satisfying scientifically, while having the intriguing potential to serve as a more accurate basis for subsurface predictions. To build a better medium for the physics, we need to incorporate knowledge from the field of petrology.

Existing rock physics models do not rigorously account for *compaction* — the reduction in bulk rock volume — despite the critical role it plays in reducing porosity in virtually all sandstone reservoirs (Lundegard 1991). In clay-poor sandstones, depositional intergranular volumes (IGVs), or 'critical porosities' in rock physics speak, generally range from 35 to 45 volume percent. Deeply buried sandstone reservoirs, on the other hand, typically have IGVs on the order of 25 percent in rocks rich in rigid grains (Paxton et al. 2002) or down to 10 percent or less in rocks with abundant lithic fragments (Pittman and Larese 1991).

The primary driving forces for compaction in sandstones are elastic deformation, plastic deformation, grain fracturing, and dissolution at grain contacts. Simulations made with our 3D grain-scale petrology model, which considers each of these compaction processes while also using realistic grain shapes derived from micro CT scans of natural sands, indicate that plastic deformation and contact dissolution generally have the largest impact on contact properties. Elastic deformation, in contrast, is much less important despite having received the most attention in existing rock physics models. Our petrology models show that although the average 'coordination number' — the number of contacts a grain has — increases somewhat during compaction, the areas of individual contacts increase greatly. Contact areas and stresses also are influenced by grain shapes and physical properties as well as the grain size distribution and vary considerably from one grain to the next.

The other critical process that should be accounted for when depicting the 'rock' is *cementation*. Cements, even in small volumes, can greatly increase acoustic velocities.

*To build a better medium for the physics, we need to
incorporate knowledge from the field of petrology.*

Quartz cement has rightly received the most attention in existing rock physics models. However, there are problems with the way it has been depicted in previous work leading to three misconceptions:

1. **It forms on all grains.** From petrology we know that quartz cement in sandstones generally forms as overgrowths where the cement represents continued growth on a pre-existing quartz crystal substrate. Non-quartz grains — a significant fraction of most reservoirs — therefore do not develop quartz overgrowths.

2. **It grows uniformly on grains or more rapidly at grain contacts or adjacent to large pore bodies.** There is considerable anisotropy in overgrowth thickness that reflects crystallographic orientations and nucleation surface types. Grain contacts and pore bodies affect cement distribution only by acting as passive barriers or reservoirs for crystal growth.

3. **It grows in uncompacted grain packs (IGV > 35%).** The thermal exposure required for quartz cement to reach measurable volumes usually is achieved only after reaching burial depths sufficient to greatly reduce IGV from compaction.

The good news is that petrology models have been developed that accurately depict quartz cement at the grain scale (e.g., Lander et al. 2008; Wendler et al. 2015).

References

Lander, R, R Larese, and L Bonnell (2008). Toward more accurate quartz cement models: The importance of euhedral versus noneuhedral growth rates. *AAPG Bulletin* **92**, 1537–1563. DOI: 10.1306/07160808037.

Lundegard, P (1991). Sandstone porosity loss — a 'big picture' view of the importance of compaction. *Journal of Sedimentary Petrology* **62**, 250–260. DOI: 10.1306/D42678D4-2B26-11D7-8648000102C1865D.

Paxton, S, J Szabo, J Ajdukiewicz, and R Klimentidis (2002). Construction of an intergranular volume compaction curve for evaluating and predicting compaction and porosity loss in rigid-grain sandstone reservoirs. *AAPG Bulletin* **86**, 2047–2067. DOI: 10.1306/61EEDDFA-173E-11D7-8645000102C1865D.

Pittman, E and R Larese (1991). Compaction of lithic sands: experimental results and applications. *AAPG Bulletin* **75**, 1279–1299.

Wendler, F, A Okamoto, and P Blum (2015). Phase-field modeling of epitaxial growth of polycrystalline quartz veins in hydrothermal experiments. *Geofluids* **16**, 211–230. DOI: 10.1111/gfl.12144.

The rock physics bridge

Zakir Hossain

Rock physics describes the relationship between rock properties and seismic properties — it is a bridge. Rock physics models help us to build up this bridge, and help us describe both lithologies and fluids. The question is how to build such a bridge? Well, we start with a foundation and a strong platform on which to build.

Here's an example of how to construct a rock physics bridge in a carbonate reservoir.

The first step is to calibrate geological depth trends for limestone (Ⓛ🌑) and dolomite (Ⓓ🌑). Then we use a petrophysics model to construct additional rocks — a dense limestone micrite (Ⓜ🌑) and a porous framestone (Ⓕ🌑) — from depth trends Ⓛ🌑 and Ⓓ🌑. Geological depth trends are the only inputs, and form the foundations of our bridge.

The next step is to construct the petrophysics platform from the petrophysical evaluation of limestone composition.

Finally the geophysicist uses rock physics to describe the relationship between porosity and velocity, constrained by local geology and petrophysics, to build the bridge. The rock physics relationships are described by Ⓓ🌑 for compacted dolomite, Ⓛ🌑 for compacted limestone, ⓁⓂ for calcite cementation, 🌑Ⓜ for compacted calcite cementation, ⒹⓁ for laminated limestone, and 🌑🌑 for compacted laminated limestone. Surprisingly, using modified Voigt–Reuss bounds is enough to describe these relationships. This is possible because geology is used as a foundation, and petrophysics is used as a platform to construct necessary inputs, with a depth-dependent petrophysics model describing the effects of burial depth and diagenesis.

Geology is used as a foundation [for our bridge],
and petrophysics is used as a platform
to construct necessary inputs.

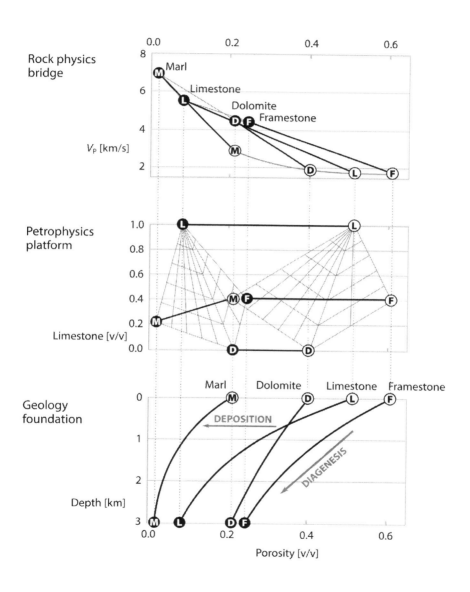

Rock physics
bridge

V_P [km/s]

Petrophysics
platform

Limestone [v/v]

Geology
foundation

Depth [km]

Porosity [v/v]

99

Three questions to ask before you start

Leo Brown

The answers to the following three questions are a good predictor of the success of a rock physics/seismic analysis project:

1. **Are the rocks favourable?** Is the physical phenomenon of interest manifested by a change in a measurable elastic property? To answer this, look at the rocks at a smaller scale than the seismic: petrophysical logs or core scale. From sonic, shear sonic, and density logs we can compute a multitude of elastic properties. Screening crossplots of the log's elastic attributes shows which of them best separate the rock property of interest from the background, and by how much. It may be that $\lambda\rho$ or $\mu\rho$ are good lithology discriminators and acoustic impedance responds to porosity changes. But lithologies may overlap in acoustic impedance, Poisson's ratio is often insensitive to porosity, fluid sensitivity usually changes significantly with depth, and permeability may not be directly predictable. Rock physics helps us understand why.

   Specific rock physics models can be perturbed for porosity, lithology, pressure, grain geometry, fracturing, or cementation. We can't evaluate all these properties simultaneously with limited petrophysical or seismic data, so the key is to find the parameter with a first-order elastic response, separable from secondary effects. If that is the property of interest, the rocks are favourable. If not, then at least you have a better understanding of the limitations.

2. **Is the stratigraphy favourable?** To answer this, we investigate rocks at the seismic scale, incorporating the wavelet and layer geometry. Since reflections are generated by the contrast in elastic properties at layer boundaries, the overburden and underburden 'background' also affect the 'reservoir' response. Reservoirs with spatially stable backgrounds are more favourable for quantitative interpretation of the reservoir.

   The variation in reservoir thickness is also critical due to the limited vertical resolution of the seismic wavelet. Below the ¼ wavelength tuning criterion (Kallweit and Wood 1982), both reservoir thickness and rock properties have a combined and inseparable effect on seismic amplitudes. This limit is typically 10–50 m or more, depending on rock velocity and seismic bandwidth. The reflection amplitude of a thin bed decreases steadily below the tuning thickness, so there is a minimum thickness below which layers cannot be detected or interpreted quantitatively. This detection thickness may be in the range of $\frac{1}{20}$ to $\frac{1}{10}$ wavelength depending on the background amplitudes and noise level. It is common that a reservoir looks amenable to rock physics analysis at the log scale, but is challenged at the seismic scale.

3. **Is the seismic data good enough?** Do we have the appropriate data type and is the fidelity, resolution, and noise level good enough? A 'no' is not necessarily fatal to the project, as data can be reprocessed, analyzed further (i.e. AVO analysis or inversion), or reacquired. The data quality needed depends on the rock property of interest identified in the log scale analysis. For rocks with low acoustic impedance and Poisson's ratio relative to background, the class 3 AVO behaviour may be anomalous on full stack as well as partial angle stack data. Reservoirs with small or even slightly positive acoustic impedance contrast but low Poisson's ratio (AVO class 2 and 1) require partial angle stacks or pre-stack AVO analysis. Some rock properties may be more demanding of acquisition than others — for example, long offsets for larger incidence angles, and close attention to amplitude-preserving processing. Anisotropy analysis requires azimuthal coverage or multi-component data.

There is generally an increase in seismic data and well calibration requirements in the progression from full stack through partial angle stack, AVO analysis, post-stack inversion, and pre-stack inversion products. The most useful measure of seismic fidelity for rock physics analysis is the quality of the well-to-seismic ties in both the pre- and post-stack domain (see Newrick 2014 for more on this topic). Only when seismic data is calibrated to log properties at known control points can it be interpreted quantitatively away from wells.

References

Kallweit, R and L Wood (1982). The limits of resolution of zero-phase wavelets. *Geophysics* **47** (7), 1035-1046. DOI: 10.1190/1.1441376.

Newrick, R (2014). Where did the data come from? In: *52 Things you should know about Geophysics.* Mahone Bay: Agile Libre.

Understand the shales to understand the reservoir

John Logel

Shales are made up of clay minerals and other clay-sized particles. They are the most prevalent deposits on earth and until the shale gas revolution, the most under-studied, under-sampled, and over-simplified. They can transform and change depending on the earth forces acting on them. Shales can be over-pressured, fractured, and diagenetically altered; they can be the source of hydrocarbons and can also be the seal that makes trapping them and exploiting them possible. Most of the world's great fossil finds are in shales. These reasons alone make them interesting and worthy of greater study.

Despite all our diverse knowledge about shales, we still myopically concentrate most of our prediction efforts on the sandstones and limestones of the reservoir rock we are evaluating. Conventional reservoir rocks are relatively easy to understand and characterize. They are dominantly made up of grains of a specific and uniform density (commonly 2650–2710 kg/m^3), specific and uniform particle size, and some pore space that is either filled with brine or hydrocarbon.

Shales and clay minerals, on the other hand, vary drastically in their occurrence and properties. They can have a wide range of particle sizes and grain densities. They can be quite variable in mineral composition, sorting and percentage. Not only do the different mineralogies vary substantially, but individual minerals themselves exhibit large distributions. Clay mineral grain densities can vary from as low as 1700 to as high as 2950 kg/m^3. Their corresponding velocities and elastic moduli have similarly wide distributions.

Rock physics is the understanding of how these rocks combine, the acoustic and elastic properties of these combinations, and the contrast with our prospective or discovered reservoir rock. It is this contrast that we are trying to map or predict. In classic geophysical modelling we examine how the rocks in our area of interest undergo compaction with depth. We know in most cases that clays and porous rocks compact at different rates. The clays characteristically compact more slowly. This is because mechanical processes prevent the non-uniform grains from easily packing together — the platy nature of the minerals traps bound water leading to very low permeability and slow rates of water expulsion. The variation of these compaction rates is a dominant component of the now

Reservoir rocks usually respond in a predictable manner,
but the 'background trend' rotates and changes with
depth, geographic location, and geological age.

famous Rutherford and Williams AVO classification. Ironically, porous reservoir rocks compact at very similar rates around the world, but the background shales vary drastically, depending on their composition.

Where the porous rock and the shale compaction curves cross marks the depth in a basin where we move from low impedance reservoir ('soft' sands) to zero impedance and eventually high impedance ('hard' sands). It becomes apparent that the standard deviation of the shale line is quite large while the porous rock standard deviation is quite small and predictable.

It is easy to conclude that the properties of a porous rock of specific grain type at a normally compacted depth are relatively easy to predict. We can with some certainty estimate its rock, acoustic, and elastic properties. However, the key to understanding its seismic response is in also understanding the properties of the surrounding background rock, most commonly shales. In crossplot space, reservoir rocks usually respond in a predictable manner, but the 'background trend' rotates and changes with depth, geographic location, and geological age.

So spend some time understanding your background rocks, look at their mineralogy, provenance, burial history, and framework. Look at the cuttings, cores, and special core analysis. The benefit of understanding these clay rocks and their relationship to your reservoir is a more accurately characterized seismic response.

What is brittleness?

Matt Hall

Brittleness is impossible to define formally because there are so many different ways of looking at it. For this reason, Tiryaki (2006) suggested calling it a rock behaviour, not a rock property, so we should try to get beyond it and ask what really matters. Mining engineers are concerned with a property called *cuttability*. Conceptually this is analogous to something we are interested in: *frackability*.

What is brittleness *not*?

- It's not the same as frackability or rock strength.
- It's not a simple rock property like, say, density.
- It's not proportional to any elastic moduli or linear combinations of them.

So what is it then?

It depends a bit what you care about. How the rock deforms under stress? How much energy it takes to break it? What happens when it breaks? Hucka and Das (1974) rounded up lots of ways of looking at it. Here are a few:

- Brittle rocks undergo little to no permanent deformation before failure.
- Brittle rocks undergo little or no ductile deformation past the yield point.
- Brittle rocks absorb relatively little energy before fracturing.
- Brittle rocks have a strong tendency to fracture under stress.
- Brittle rocks break with a high ratio of fine to coarse fragments.

All of this is only made more complicated by the fact that there are lots of kinds of stress: compression, tension, shear, torsion, bending, and impact… and all of these can operate in multiple dimensions, and on multiple time scales.

Some brittleness indices

There are lots of approaches to brittleness in the literature. Several of them capture the relationship between compressive and tensile strength, σ_C and σ_T

respectively. This is potentially useful, because we measure uniaxial compressive strength in the standard triaxial rig tests that have become routine in shale studies. But the tensile strength is much harder to measure. This is unfortunate, because hydraulic fracturing is initially a tensile failure (though reactivation and other failure modes do occur — see Williams-Stroud et al. 2012).

Altindag (2003) gave the following examples of brittleness indices — the strength ratio, a strength contrast, and the mean strength (his favourite):

$$B = \frac{\text{compressive strength}}{\text{tensile strength}} = \frac{\sigma_C}{\sigma_T} \qquad B = \frac{\sigma_C - \sigma_T}{\sigma_C + \sigma_T} \qquad B = \frac{\sigma_C + \sigma_T}{2}$$

This is just the start, once you start digging, you'll find lots of others. Like Hucka and Das's definitions above, one thing they have in common is that they capture some characteristic of rock *failure* and do not rely on implicit rock properties.

What to do

The prevailing view among many interpreters is that brittleness is proportional to Young's modulus and/or Poisson's ratio, and/or a linear combination of these. Lev Vernik said at the SEG Annual Meeting in 2012 that we need to realize that the rocks we'd like to frack are not isotropic, and that computing shale brittleness from elastic properties is not physically meaningful. For one thing, you'll note that elastic moduli don't have anything to do with rock failure.

Where does that leave us? Here's Altindag (2003) again:

> Brittleness, defined differently from author to author, is an important mechanical property of rocks, but there is no universally accepted brittleness concept or measurement method...

We should stop worrying about brittleness, whatever it is, and focus on things we really care about, like organic matter content or frackability. The thing is to collect good data, examine it carefully with proper tools (Spotfire, Tableau, R, Python...), and find relationships you can use — and prove — in your rocks.

References

Versions of this essay first appeared as blog posts in 2013: *ageo.co/whatbrit* and *ageo.co/whichbrit*

Altindag, R (2003). Correlation of specific energy with rock brittleness concepts on rock cutting. *The Journal of The South African Institute of Mining and Metallurgy* **103** (3). *ageo.co/23XLirv*

Hucka, V and B Das (1974). Brittleness determination of rocks by different methods. *Int J Rock Mech Min Sci Geomech Abstr* **10** (11), 389–92. DOI:10.1016/0148-9062(74)91109-7.

Tiryaki, B (2006). Evaluation of the indirect measures of rock brittleness and fracture toughness in rock cutting. *The Journal of The South African Institute of Mining and Metallurgy* **106** (6), June 2006. *ageo.co/1XAYUsZ*

Williams-Stroud, S, W Barker, and K Smith (2012). Induced hydraulic fractures or reactivated natural fractures? Modeling the response of natural fracture networks to stimulation treatments. American Rock Mechanics Association 12–667.

What is Gassmann's equation?

Ali Misaghi

Seismic methods are applicable for the entire lifecycle of a reservoir, not only for the exploration phase. This is especially true for sandstone reservoirs. Seismic reservoir monitoring is now an established production monitoring tool; we're most interested in four types of reservoir changes:

1. Saturation changes.
2. Pressure changes.
3. Geomechanical changes.
4. Temperature changes.

The question for geophysicists is how these physical changes can be related to changes in seismic properties of the medium. The answer is often given by rock physics models — Gassmann's model is probably the most well known. Along with Biot theory, Gassmann theory is used to predict how P- and S-wave velocities change as saturation changes. The underlying assumptions are discussed by White (1983), Bourbié et al. (1987), and Mavko et al. (1998). Gassmann's theory (Gassmann 1951) applies to the low-frequency case and is identical to the low-frequency limit of Biot's theory (Biot 1956a) for the high-frequency range. Biot (1956b) also derived theoretical expressions for predicting P- and S-wave velocities of saturated rocks in terms of dry rocks. For the case of an isotropic, liquid-saturated porous medium Biot's theory predicts two coupled P-waves and one S-wave.

There are other rock physics models for estimation of seismic velocity versus saturation. Gassmann's model is based on some specific assumptions, so the concordance of rocks and fluid properties with these assumptions should be evaluated before using it. The other theories include effective medium theories, self-consistent theories, and differential effective medium theories.

Gassmann's model

The most widely-used theory for fluid substitution is the low-frequency Gassmann theory. Gassmann's equation gives a relationship between saturated bulk modulus, porosity, bulk modulus of rock frame, bulk modulus of minerals of rock matrix and the bulk modulus of pore fluids (Mavko et al. 1998):

$$K_{sat} = K_{dry} + \frac{\left(1 - \dfrac{K_{dry}}{K_m}\right)^2}{\dfrac{\varphi}{K_{sat}} + \dfrac{1 - \varphi}{K_m} + \dfrac{K_{dry}}{K_m^2}}$$

where, K_{sat} is the saturated bulk modulus, K_{dry} is the bulk modulus of rock frame, K_m is the bulk modulus of minerals of the rock matrix, K_f is the bulk modulus of the pore fluids and φ is the porosity.

Gassmann's theory makes several assumptions, which must be taken into account in any application (Wang 2001):

1. Rock (matrix and frame) must be macroscopically homogeneous.

2. All pores must be interconnected.

3. Pores are filled with a frictionless fluid.

4. The rock-fluid system must be closed (undrained).

5. There should be no interaction between fluid and the matrix in a way that could soften or harden the frame.

The first assumption implies that the wavelength must be greater than the pore and grain sizes. The second assumption indicates that the porosity and the permeability must be high and there should be no isolated or poorly connected pores. Assumptions 2 and 3 explain why wireline and laboratory velocity data are usually higher than Gassmann's predictions. For some frequencies, a relative movement between the fluid and the matrix will occur, and this might lead to dispersive waves. Relative fluid–matrix motion is generally more prominent for some special frequencies — a sort of resonance — and might create large differences between bulk and shear moduli of fluid and matrix.

With that equation and those assumptions, saturated bulk modulus K_{sat} can be estimated. Once we know K_{sat}, P- and S-wave velocities V can easily be predicted (Mavko et al. 1998), given shear modulus μ and bulk density ρ:

$$V_P = \sqrt{\frac{K_{sat} + \frac{4}{3}\mu}{\rho}} \qquad V_S = \sqrt{\frac{\mu}{\rho}}$$

References

Bourbié, T, O Coussy, and B Zinszner (1987). *Acoustics of Porous Media*. Paris: Editions Technip.

Gassmann, F (1951). Über die Elastizität poröser Medien. In: *Vierteljahrsschrift der Naturforschenden Gesellschaft in Zürich*. Band 96, 1951, S. 1–23. Available in English at *ageo.co/gassmann51*.

Biot, MA (1956). Theory of propagation of elastic waves in a fluid-saturated porous solid. Part I: Low-frequency range and Part II: Higher-frequency range. *The Journal of the Acoustical Society of America* **28** (2). Part I: 168–178, Part II: 179–191.

Mavko, G, T Mukerji, and J Dvorkin (1998). *The Rock Physics Handbook – Tools for Seismic Analysis in Porous Media*. Cambridge University Press, UK.

Wang, Z (2001). Fundamentals of seismic rock physics. *Geophysics* **66** (2), 398. DOI: 0.1190/1.1444931.

White, J E (1983). *Underground Sound — Applications of Seismic Waves*. New York: Elsevier.

What is the speed of sound?

Dave Monk

Why can't a geophysicist give a straight answer to the question, 'What is the speed of sound in the rock?'? You would think this trivial but important question would have a simple answer. Surely velocity is a fundamental rock property? Surely it can be determined?

We need to know the velocity in order to do depth conversion of seismic data; we need to know the velocity in order to compute the acoustic impedance of a material (assuming we also know the density); and we need impedance to compute the reflection coefficient between two rocks. So it's important. Why is it complicated?

Firstly let's limit the discussion to simple P waves — compressional sound waves in which the direction of motion of the atoms in the medium is in the same direction as the propagation of the sound wave through the medium. The velocity determined through the study of seismic data from the surface may have little or nothing to do with the velocity measured using a down-hole tool. Velocity from seismic data is typically determined along a path through the earth which is not easily determined, and certainly not in a single direction. On the other hand, the down-hole measurement is usually made in rock very close to the borehole, and typically in a direction parallel to it. While it may be reasonable to assume that there is a single velocity involved in a single down-hole measurement (perhaps over a few centimetres of rock), seismic velocities are typically derived over a path length of several kilometres, and the velocity is far from constant along the path. Under some special simplifying circumstances it might be possible to determine a velocity along the ray path, but this will be close to the root-mean-square velocity, and not (for example) the average vertical velocity which might be desired for depth conversion of seismic data. But the earth is far more complicated than that.

The speed of sound through a sample of rock is firstly dependent on the direction of propagation of the sound. It is usually different in a vertical plane compared to a horizontal plane (vertical anisotropy), and often varies azimuthally in the horizontal plane as well (azimuthal or horizontal anisotropy).

All rocks attenuate the energy associated with sound as it passes through the

The velocity determined through the study of seismic data
from the surface may have little or nothing to do with
the velocity measured using a down-hole tool.

rock matrix (sometimes referred to as the Q effect, or acoustic attenuation), as some of the energy is transferred to heat. This attenuation is dependent on the frequency of the sound, and leads to a slowing of the sound as it passes through the rock. The higher the frequency, the more the sound is attenuated. But this effect is also dispersive, which means that each frequency must also be travelling at a different speed through the rock. So when a geophysicist performs Q correction on seismic data, he picks a 'reference frequency' which will be the frequency that is corrected back with an appropriate velocity. Think about what this means for log versus seismic velocity estimation when velocity measurements from down-hole tools typically use frequencies in the 10s of kHz, whereas seismic data may peak at well below 100 Hz. An interesting corollary to this is that it means that the reflection coefficient between two rocks is not only reflection angle dependent, but also frequency dependent.

I started this essay by suggesting I would limit the discussion to P waves, but of course there are other modes of propagation of sound through a rock matrix. The most energetic of these other modes is commonly referred to as S or shear waves. This introduces a whole new set of complicating factors, which make the simple question I started with far more complicated than this short article will allow. So please forgive the geophysicist, not only does he not know the answer; he's not even sure what the question actually is.

What use is AVO?

Duncan Emsley

Seismic amplitude variation with offset (AVO) can be an important reservoir characterisation tool. The AVO character of reflective interfaces between geological units is related to the change in acoustic impedance and V_P/V_S ratio across these interfaces. Every reflector has an AVO response — what we are looking for are anomalous AVO responses against the background trend that can be mapped and related to some useful geological change in porosity, fluid, sand quality, etc.

Amplitude behaviour with offset for P and S data are described by the Zoeppritz (1919) equations which describe how compressional (P) and shear (S) waves interchange at interfaces. For the P-wave case, which forms the majority of seismic exploration datasets, the Zoeppritz equations, which are quite complicated, can be approximated by Shuey's two-term equation:

$$R(\theta) = A + B \sin^2\theta$$

where A is the zero-offset reflectivity governed by the change in acoustic impedance across the interface, and B is the AVO gradient term which is dependent upon the change in V_P/V_S ratio. In amplitude versus $\sin^2\theta$ space, this is the equation of a straight line (A is the intercept and B is the slope) and is applicable up to incidence angles of 30–40 degrees. For offsets beyond this, higher order equations are available which include some dependence upon density contrast.

Geological changes such as shale to sand, lateral changes in porosity or sand quality, and updip changes in fluid each have their own influence on the acoustic impedance and V_P/V_S. These will then drive an AVO response across a geological interface. The trick is to identify and describe these changes and infer their geological causes.

A common way to analyze AVO is to crossplot A and B from Shuey's equation, after Foster et al. (2010) — see figure opposite. The AVO classification shows a rotational symmetry with the top of sands plotting to the lower left of the shale line for SEG normal polarity. Base sands plot above and to the right. Classes serve as a convenient way to characterize an interface, but they are not definitive. The AVO crossplot is commonly normalized for the background shale response. Wet sands tend to lie close to (or overlapping) the shale trend — softer fluids plot orthogonally away from this trend. Different reservoir characteristics

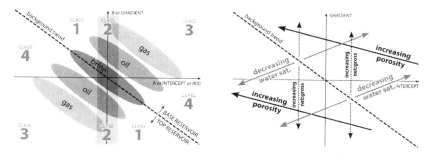

affect the locus of points according to the vectors shown. Porosity variations plot sub-parallel to the shale trend with higher porosity giving a softer (class 3 or 4) response. Net:gross is predominantly a gradient effect.

The AVO method of extracting zero offset amplitude (intercept) and the change in amplitude with offset or angle (gradient) requires having correctly processed common depth point or common image point gathers. Moveout correction should be handled with care as the properly corrected gathers may not always appear to be completely flat (Emsley 2012). In seismic sections, AVO anomalies can appear as a gradual phase change in certain circumstances and this can confuse a velocity picker — automated or human. Any tool in the processing sequence that is likely to change the waveform or amplitude with offset should be designed and applied to the ensemble as a whole and never trace by trace.

Geology, too, can be a pitfall because it is as likely to change above an interface as it is below, both with the chance of enhancing or reducing the AVO response. So it is best, if possible, to commence any study with some work on expectations.

Sufficient source-receiver offset is required to produce well-sampled angles of incidence of the order of 30°–45°. Higher angles *may* allow the derivation of density from the AVO response, but this requires a great deal of accuracy in flattening the gathers. Increasing depth and velocity will increase the angle of incidence and may make AVO less effective if these angles are not sufficiently sampled in acquisition. Pre-stack inversion uses precisely the same seismic inputs as AVO — flattened gathers. But it also requires well information for wavelet estimation and for building the low frequency model. AVO may therefore appear to lend itself to a more exploration-oriented area where well data is hard to come by. However, it still has its place in development as acoustic impedance and V_P/V_S changes are combined in a single analysis.

References

Emsley, D (2012). Know your processing flow. In: *52 Things you should know about geophysics*. Mahone Bay: Agile Libre.

Foster, D, R Keys, and F Lane (2010). Interpretation of AVO anomalies. *Geophysics* **75** (5), 75A3–75A13. DOI: 10.1190/1.3467825.

Shuey, R T (1985). A simplification of the Zoeppritz equations. *Geophysics* **50**, 609–614. DOI: 10.1190/1.1441936.

Zoeppritz, K (1919). Erdbebenwellen VIII B, Über die Reflexion und Durchgang seismischer Wellen durch Unstetigkeitsflächen. *Gottinger Nachr* **1**, 66–84.

Where did you get those logs?

Rachel Newrick

Just as it is possible to interpret seismic data without having studied acquisition and processing, it is possible to interpret and use well logs without an understanding of the acquisition and processing — but an understanding of the origin of the data allows you to recognize some basic pitfalls. Here is some advice:

1. **Visit the logging contractor's operations centre** and inspect the tools. Ask questions about the history of the tools, how they were developed over time, and about new developments.

2. **Get in the field as often as possible.** Sit with the logging crew. Record observations and ask questions. Be prepared to get up in the middle of the night to see the zone of interest logging in real time.

3. **Request a copy of the tool string schematic** and watch it being assembled into the hole. Tick off each section as it is added.

4. **Request a copy of the time–depth log.** Ideally you will see the tools run into the hole and then pulled slowly out at the recommended speed. But what you may see is a mountainous profile illustrating that the tools ran in only partly, got stuck on a ledge, were pulled up, lowered slightly, stuck, up, down, finally at the bottom of the hole, pulled up, stuck, released, pulled up, lowered to re-record the section missed (irradiating the hole with the neutron tool so that gamma-ray reads high on the next pass), pulled up faster than recommended to minimize the chances of getting stuck again, skipped a section, and finished with a good section at the correct speed.

5. **Record logs on the way into the hole** — if you lose the string, you'll have some data. On key exploration wells, logging while drilling is recommended. On a high-cost frontier well we failed to get wireline data over an interesting seismic anomaly due to hole washout.

6. **Be friendly with the wellsite geologist.** They are often in charge of log quality and can alert you to any concerns. An experienced wellsite geologist has also seen many well logs and can help with the initial interpretation of the data, especially when there are interesting or perplexing anomalies.

7. **Ask to see raw curves.** Remember that well logs are processed to obtain the logs you interpret. Many recordings and parameters are needed, so run

through them with your logging contractor. Ask how the final curves were derived. Question anything that looks odd.

8. **Archive field data (DLIS files)** so that you can reprocess the logs in the future.

9. **Get a copy of your contractor's log interpretation charts.** For specific curves and tools, ask to see any white papers or internal documents that discuss the tool.

10. **Know the borehole environment:** drilling mud, circulation times, additives, casing points, logging runs. A hydrocarbon show may be a change from water-based to oil-based mud. A gamma-ray increase may be a decrease in borehole size so the tool is closer to the formation, without the appropriate environmental correction.

11. **Never use the logs in isolation** — plotting the full suite will help identify zones of interest and concern. For example, a large zone of washout, identified by a high caliper reading, will render density unusable. Density is a pad tool (you recall from your visit to the contractor's operational centre and again in the field), so if there is no contact with the formation there is no usable density measurement. Consequently density is often derived from other curves using observed relationships, but our interest is in the anomalous zones so this isn't particularly useful for interpretation.

12. **Create an 'origin curve' for the logs.** This provides a clear record as to whether a curve was recorded on wireline, logging while drilling (LWD), or created from other curves to fill in a data gap.

13. **If it looks odd, question it.** This is worth repeating! In exploration, we are looking for anomalies. Do your best to ensure that interpreted anomalies are representative of genuine changes in the subsurface by understanding the acquisition, processing, splicing, dicing, and presentation of well logs.

14. **Spend time working with a petrophysicist or well-logging specialist.** Where you can't get into the field, or don't have time to visit acquisition or processing contractors, their experience can help you learn and their skills can help you obtain a superior logging suite, so take every advantage of them!

Why do we need rock physics?

Marc Sbar

Rock physics is what makes seismic interpretation possible. Every prediction depends on our knowledge of how rocks behave when seismic waves reflect from them or pass through them. Let's consider an example.

When I first interpreted seismic in the early 1980s it was on the Morrow play in Oklahoma. The Morrow sandstones discovered so far in that area were low porosity and high impedance, so the signature of pay was a peak/trough event. The sand was also thin enough to be below tuning, so that a low amplitude event would be thin and a higher amplitude peak/trough would be thicker. My partner and I were poring over the 2D data (on paper in those days) and we saw a low amplitude zone along the Morrow horizon that appeared to be thicker in time than usual. This was a substantially different response from the one the team was expecting. With our basic understanding of how sandstones behave from rock physics studies, we were able to predict that this might be a higher-porosity, above-tuning event. After convincing our management of this, the most productive well found by the team up to that time was drilled, verifying the rock physics part of the prediction.

When geophysicists began to use seismic in earnest they only had stacked data. Amplitude on the stacked section was the key to guessing where hydrocarbon-filled reservoirs existed. But this can only take you so far. Essentially one parameter, vertical contrast in acoustic impedance, can be predicted from stacked seismic data. In some reservoirs this is sufficient to distinguish pay from non-pay, but as the simple reservoirs were developed more information was needed to constrain the various reservoir properties.

Amplitude versus offset (or, equivalently, angle) analysis evolved to provide that additional information. Using pre-stack seismic data, amplitude variation is measured as a function of angle of incidence and predictions of the contrast for *two* elastic properties can be made. The elastic properties extracted via this kind of analysis might be acoustic impedance and Poisson's ratio, or $\lambda\rho$ and $\mu\rho$ for example. Other pairs of properties can be computed from these.

From these elastic properties, and a knowledge of rock physics, reasonable predictions of porosity, lithology, and fluid content can be made. Since there are other relevant reservoir properties in addition to these, the problem is not well constrained. It is necessary to bring all of the geological knowledge for a prospect to bear in order to constrain the problem and make reliable predictions.

Where there is some well control, absolute rather than relative values of reservoir properties can be made. Amplitude-vs-offset analysis and two-term inversion took exploration and development to the next level, but there are some environments where we need still more discrimination. Where there is even more well control, inversion with long-offset, high-quality seismic data can extract three elastic properties (say, density, V_P, and V_S).

Often rock physics relationships between reservoir properties and elastic properties, such as acoustic impedance or Poisson's ratio, are based on generic relationships derived from a large cohort of different rocks from many different environments. It is always better, however, to obtain measurements of the specific rocks from the reservoir of interest to ensure the most accurate relationships.

In the end we are interested in estimating reservoir properties not elastic properties. Rock physics is the connection between the two. It is what makes seismic interpretation possible.

Why echoes fade

Ludmila Adam

Waves surround us in our daily lives, but already as young children we learn that the behaviour of waves depends on the setting. A child's ears under water in the bath cannot hear their parent's voice. Likewise, as we experience the full range of frequencies in our favourite music, our neighbours on the other side of the wall may not enjoy the arrival of only the low bass frequencies, as the higher-pitched treble sounds are lost.

There are several reasons why the person in the next room only hears the lower tones in the music. The disproportionate loss of high frequencies is due to a combination of effects: wave scattering, absorption, and geometrical spreading.

1. **Scattering.** Walls have a much higher density than air. The speed of sound in a wall is also higher than in air. The product of these properties, the acoustic impedance, determines the ratio of reflected versus transmitted sound. It may be intuitive to the reader that a thicker wall forms a larger obstacle for sound waves than a thinner wall. For music, we define 'thickness' relative to the wavelength of the sound. For a given wall, higher frequencies are more reflected, and therefore less transmitted. This is what we refer to as wave energy loss or attenuation due to scattering of waves.

2. **Absorption.** Sound waves propagate by compressing and dilating the air in which they travel. However, when there is movement there is friction, and kinematic energy is lost to heat. In other words, wave energy is absorbed. Since high frequency waves move more air molecules in a given space than low frequency waves, losses due to friction are greater for high frequencies.

3. **Geometrical spreading.** There are also losses in sound-wave strength due to geometrical spreading, where a fixed amount of energy is spread over an expanding wave front. Assuming the expanding wave front is spherical, the area of this sphere increases and the fixed wave energy is distributed over the growing area, decreasing the wave amplitude. However, this effect is not frequency dependent.

Seismic wave energy losses can be similarly explained, but rather than playing music in our house, seismic waves are created by impulsive ground deformation. Instead of walls, we have layers in the earth. Subsurface fluids set in motion by the seismic waves are mostly responsible for the absorption mechanisms in the earth, while waves reflecting from layers or heterogeneous bodies with large impedance contrasts result in scattering losses.

In exploration geophysics, the observed wave absorption mostly depends on the pore fluid (e.g. oil, water, carbon dioxide, or fluid combinations) and rock permeability. Permeability has been one of the most elusive physical properties in the subsurface, one that geophysicists are still trying to extract quantitatively from geophysical data. Rock physics models aim to correlate rock physical properties to wave attenuation, but not all are validated by data. However, some of the models include the rock-fluid mechanisms that control seismic wave attenuation and dispersion (i.e. frequency-dependent velocity) and have been supported by experiments and field observations. Laboratory experiments can be time consuming and therefore limited in the range of rock types that can be studied. At a larger scale, wave attenuation in sonic logs is matched to the rock physical properties, but is typically done on a field-by-field basis. A major challenge here is separating scattering from intrinsic attenuation, which can be tackled by numerical modelling of the scattered wavefield. Stoneley wave analysis to extract quantitative estimates of wave absorption from full-waveform sonic logs is an advancing technology which might become standard practice.

The most spatially extensive data with which to estimate quantitative absorption information is surface seismic data. Again, however, geometric effects and noise affect the estimate of seismic attenuation and dispersion. Few studies provide a comprehensive analysis of seismic wave absorption by integrating core, log, and seismic data. Although theoretical models, laboratory measurements, and field observation have advanced our understanding of seismic wave amplitudes, a major challenge ahead is correlating the differences in spatial and frequency scales, and separating scattering attenuation from absorption.

List of contributors

Ludmila Adam is a senior lecturer at the University of Auckland, New Zealand. She has held positions at Boise State University, Colorado School of Mines, and Sincor. She has an MSc and a PhD in geophysics from the Colorado School of Mines.

Why echoes fade 116

Paul Anderson is a senior geophysicist and researcher with Denbury Onshore LLC in Plano, Texas. His background is primarily in prestack inversion, AVO, rock physics, microseismic, multicomponent seismic, 4D seismic, and data processing. Paul received his BSc in geophysics in 1998 and his MSc in 2010, both from the University of Calgary. Prior to his current work at Denbury focusing on 4D seismic data analysis and geophysical research, Paul has worked at Apache Corporation, Veritas Geoservices, and Hydro-Fax Resources.

Pitfalls of anisotropy 66

Per Avseth is a geophysical adviser and researcher at Tullow Oil Norge in Oslo, Norway. He is also an Adjunct Professor in applied geophysics at the Norwegian University of Science and Technology (NTNU) in Trondheim, Norway. His research focus is on quantitative seismic interpretation and rock physics analysis. Per received his MSc in applied petroleum geosciences from NTNU in 1993, and his PhD in geophysics from Stanford University, California, in 2000. He was the SEG Honorary Lecturer for Europe in 2009. Per is a co-author of the book *Quantitative Seismic Interpretation* (Cambridge University Press, 2005).

Building a good low-frequency model 26

Ran Bachrach is a scientific advisor on rock physics and geophysics with Schlumberger. He holds a PhD in geophysics and PhD minor in civil engineering from Stanford University.

Right, wrong, and useful rock physics models 76

Gilles Bellefleur is a research scientist at the Geological Survey of Canada. For the last 20 years he has been working on various aspects of seismic methods applied to crystalline rocks and mineral exploration in mining camps located across Canada. In particular, he is interested in the petrophysics of massive sulphide deposits and their signature on seismic gathers and sections. Occasionally, he works on seismic data acquired in 'soft' rock environments to remember what good data looks like. He is a member of EAGE and SEG, and associate editor for *Geophysical Prospecting*.

Hard rock is (not) like soft rock 46

James G Berryman is a senior scientist at Lawrence Berkeley National Laboratory in California. Prior to 2006, he worked as a physicist at the Lawrence Livermore National Laboratory for 25 years. He received his PhD from the University of Wisconsin-Madison and a BSc from the University of Kansas.

Beware of shortcuts 24

Evan Bianco works at Agile Geoscience. He has more than 10 years of consultancy experience involving geological modelling, scientific computing, and integrating multidisciplinary data sets in the subsurface. He has an MSc in geophysics from the University of Alberta, where he studied rock physics. He has taught introductory geophysics to undergraduate students, and regularly teaches fellow geoscientists how to program. You can reach him at *evan@agilegeoscience.com*, or you can follow him on Twitter *@EvanBianco*.

Elastic symmetry: isotropy 36
Elastic symmetry: anisotropy 38

Stephen Brown is a research scientist in the Department of Earth, Atmospheric, and Planetary Sciences at the Massachusetts Institute of Technology in Cambridge, Massachusetts, and a consultant working under the auspices of Tierra Sciences Ltd in Montpelier, Vermont. His primary scientific interests are the physical properties of fractured and otherwise heterogeneous rock, as well as the application of geophysical methods to mining, energy, and environmental problems.

My box of rocks 62

Leo Brown is a geophysicist at ConocoPhillips and has worked in Oklahoma, Texas, Norway, and currently Alaska. He has enjoyed all aspects of geophysics, having worked in research at the USGS, engineering geophysics at GeoVision Geophysical Services, and seismic technology, exploration, and development within ConocoPhillips since 2002. He holds a BSc in engineering geology from BYU (1997), an MSc in geotechnical engineering from UT Austin (1998), and an MSc in geophysics from Colorado School of Mines (2002). His thesis work in CSM's Reservoir Characterization Project involved integrating rock physics with reservoir simulation for interpreting 4-D seismic data. One of the things Leo likes best about science in the petroleum industry is the opportunity to test his work with the drill bit.

Three questions to ask before you start 100

Arthur C H Cheng is a Professor in petroleum engineering at the Department of Civil and Environmental Engineering, National University of Singapore. Prior to September 2014, he was senior manager for acoustics and borehole seismics at Halliburton Technology. He received a BSc in engineering physics from Cornell University in 1973, and a ScD in geophysics from MIT in 1978. He was one of the co-founders of the Earth Resources Laboratory at MIT in 1982, and was project leader of the MIT Borehole Acoustics and Logging Consortium until 1996, when he joined Western Atlas as manager of acoustic science. He has also worked for Baker Atlas, Baker Hughes Inteq, SensorWise, and Rock Solid Images in various managerial and consulting capacities. He has published over 100 papers and 10 patents, with more pending. Arthur is currently an assistant editor for *Geophysics* in borehole geophysics and rock properties. He received the Life Membership Award from SEG in 2013 and the Distinguished Technical Achievement Award from SPWLA in 2015.

How to catch a shear wave 50

Sagnik Dasgupta is a senior rock physicist with Schlumberger where he has worked since 2006. He received a BSc in geology from the Presidency College, Kolkata, and his MSc and MTech in exploration geoscience from the Indian Institute of Technology, Bombay. Since joining Schlumberger he has been involved in applied reservoir geophysics, petrophysics, rock physics, and geomechanics for both conventional and unconventional resources. His main research focus is on the effect of rock fabric, discontinuity, and regional stresses on different geophysical measurements.

Measurements are scale dependent 56

Franck Delbecq is a senior geophysical specialist with Nexen, a CNOOC company. He was previously reservoir services manager with Hampson-Russell, a CGG company. His professional work has been focused on applied reservoir geophysics for both conventional and unconventional resources, with experience in Europe and North America. Franck received a master of reservoir engineering degree in 2005 from the Nancy School of Geology, France, and has worked in Calgary since 2007. Franck has presented multiple papers about inversion and fracture analysis and enjoys sharing ideas with the geoscience community.

Here be dragons: the need for uncertainty 48

Rocky Detomo received his PhD in physics from Ohio State University in 1981 then joined Shell Oil Company as a geoscientist. Rocky's career with Shell spanned 33 years where he served as supervisor of seismic acquisition and processing in both the onshore and offshore United States, was an offshore interpreter and project manager for the Gulf of Mexico, led global deepwater exploration capability deployment, served as an exploration seismic manager, and was head of Sub-Saharan Africa reservoir geophysics. Before retiring from Shell in 2014 and founding Omoted Geophysical Consulting, Rocky also led Shell's areal monitoring research teams. Rocky has served as president of SGS, SEG board director, SEG's annual meeting technical program chairman, SEG honorary lecturer, and serves as a trustee associate of the SEG Foundation.

Mathematical descriptions of physical phenomena 54

Dave Dewhurst is a geologist by background with a BSc from the University of Sheffield and a PhD from the University of Newcastle, both in the UK. He has worked for over 20 years on microstructure and rock properties, with an emphasis on clay and shale behaviour. He did postdoctoral stints at the University of Birmingham, University of Newcastle, L'Institut Français du Pétrole, and Imperial College investigating physical properties, compaction, faulting and fluid flow in mudrocks. Dave moved to CSIRO in 1998 where he has worked on overpressure, fault and top seals, and the links between geomechanics, rock physics, and petrophysics in shales.

Shale rock physics   78

Peter M Duncan is founder and CEO of MicroSeismic, Inc, a Houston-based oilfield service company specializing in hydraulic fracture stimulation surveillance and evaluation. He holds a PhD in geophysics from the University of Toronto. His early career as an exploration geophysicist was with Shell Canada and then Digicon Geophysical. In 1992 he was one of three founders of 3DX Technologies Inc, a publicly traded independent oil and gas exploration company. Duncan was 2003–04 president of the Society of Exploration Geophysicists (SEG). He was the fall 2008 SEG/AAPG Distinguished Lecturer speaking on the subject of passive seismic and he is an honorary member of SEG, CSEG, GSH, and EAGE. He received the Enterprise Champion Award from the *Houston Business Journal* in 2010, the World Oil Innovative Thinker Award in 2011, and the EY Energy National Entrepreneur of the Year Award for 2013. In 2014 he received the Virgil Kauffman Gold Medal from SEG.

Mapping fractures   52

Dan Ebrom works as a leading researcher for Statoil in its research and technology division. He received his PhD in geology and geophysics at the University of Houston, and his undergraduate training at MIT. The V_P/V_S subsalt pressure estimation method described in the essay won BP's Helios Innovation prize in 2005.

Pressure signals are everywhere   72

Duncan Emsley graduated with a BSc from the University of Durham in 1984. He worked for processing contractors for several years before joining Phillips Petroleum in 1992. Continuing in the seismic processing vein, he worked with data from all sectors of the North Sea and northeast Atlantic. The merger of ConocoPhillips brought about moves to Scotland, Alaska, Canada, and back to Scotland again and a progression into rock physics and seismic attributes and their uses in the interpretation process.

What use is AVO?   110

Xinding Fang received his PhD in geophysics from Massachusetts Institute of Technology in 2013 and is now a researcher at Chevron, Houston. His research interests focus on borehole geophysics, reservoir seismic characterization, and wave propagation simulation.

Sonic logs are not true   84

Michael Fehler received his PhD in geophysics from MIT in 1979 then worked as an Assistant Professor at Oregon State University for five years before becoming a staff geoscientist at Los Alamos National Laboratory in 1984. At Los Alamos, he was the leader of the geophysics group and later the division director of the Earth and Environmental Sciences Division. In 2008 he returned to MIT where he is a senior research scientist and the deputy director of ERL. In addition to his work at MIT, Fehler is technical project manager for phase I of the SEG Advanced Modeling project (SEAM) — an industry consortium for geophysical modelling. He was Editor-in-Chief of the *Bulletin of the Seismological Society of America* for nine years beginning in 1995, and president of the Seismological Society of America from 2005–07.

Sonic logs are not true

Vedad Hadziavdic is a senior geophysicist at Wintershall in Bergen, Norway. He has worked on many aspects of quantitative reservoir characterization, applied in both exploration and production. His current research interests include statistical rock physics and 4D. Vedad received his MEng in applied physics (2000) from the University of Tromsø, Norway. He did his PhD (2005) in statistical physics at the Max-Planck Institute for Extraterrestrial Physics in Munich, Germany.

Be careful when fitting models

Matt Hall is the founder of Agile Geoscience. A sedimentologist who found geophysics later in his career, Matt worked at Statoil in Stavanger, Norway, then Landmark and ConocoPhillips in Calgary, Alberta, and is now running Agile from its world headquarters in Mahone Bay, Nova Scotia. He is passionate about communicating science and technology, and especially about putting specialist knowledge into the hands of anyone who needs it. Matt read geology at the University of Durham, and has a PhD from the University of Manchester. He blogs at *agilegeoscience.com*, podcasts at *undersampledrad.io*, and tweets as *@kwinkunks*.

Wes Hamlyn is a geophysicist with a passion for quantitative interpretation and geoscience technology. While completing his studies at Memorial University, Wes worked with Norsk Hydro's Scotian Shelf exploration team performing regional seismic mapping. Upon graduation, Wes joined Paradigm Geophysical where he spent the next 10 years supporting, promoting, and ultimately managing development of Paradigm's seismic interpretation and visualization software. In 2014 Wes joined Ikon Science to focus on rock physics, seismic inversion, and quantitative interpretation. Wes lives in Calgary, Canada.

Take the time to explore relationships

Zakir Hossain is a senior rock physicist at ION Geophysical, Houston, and before that worked for Rock Solid Images. He has an MEng and PhD in geophysics from Danmarks Tekniske Universitet. He specializes in rock physics, petrophysics, rock mechanics, AVO, and seismic. His current areas of interest include the rock physics and petrophysics of unconventional and conventional reservoirs.

The rock physics bridge												98

Lee Hunt graduated from the University of Alberta with a BSc in geophysics in 1990. He currently consults for Jupiter Resources Inc, where he holds the position of senior technical advisor, geophysics. Lee was the 2011–12 CSEG Distinguished Lecturer, and is one of the founding members of the Value of Integrated Geophysics steering committee. He was a participant in the creation of the CSEG Master Seismic Data License Agreement, APEGGA's Quality Inspection Practise Standard, as well as APEGGA's Guideline for the Ethical Use of Geophysical Data. He is also one of the principal designers of the CSEG *Value of Geophysics with Case Histories* course. Lee is an Ironman triathlete, and an enthusiastic sport rock climber.

Quantitative interpretation or quantitative everything?									74

Paul Johnson works on the physical properties of materials, especially geomaterials, applying primarily acoustical methods including a technique known as Time Reversal. He also works on the frictional properties of earth faults, attempting to understand dynamic earthquake triggering as well as working on advances in earthquake forecasting applying machine learning to seismic data. Johnson has a degree in geology from the University of New Mexico, a masters in geophysics from the University of Arizona, and a doctorate in physical acoustics from the Sorbonne Université, Paris. He is a Fellow of the Acoustical Society of America, the American Geophysical Union, and the Los Alamos National Laboratory.

The astonishing case of non-linear elasticity										90

Michael King received a BSc in mechanical engineering from Glasgow University in 1953. He spent five years as a field engineer with the Iraq Petroleum Company in Iraq and Syria before completing a PhD at the University of California at Berkeley. He has held several university posts including Professor of Geological Engineering at the University of Saskatchewan, Canada, Professor of Mechanical Engineering at Lawrence Berkeley Laboratory, California, and Oil Industry Chair in Petroleum Engineering at Imperial College, London, where he lectured at the Royal School of Mines and conducted research in petrophysics and applied geophysics. He has published over 180 papers. He retired from Imperial College as Emeritus Professor in 1996, but remains active in research, lecturing, and consulting on a part-time basis as a Senior Research Fellow.

Acoustic emission												16

Rob Lander develops models of diagenetic processes as a means to predict rock properties away from well control and through geological time. He is a co-founder of Geocosm LLC where he is part of a team of scientists, mathematicians, and software engineers that develop Touchstone, Prism2D, Cyberstone, and several other systems. He co-leads the Consortium for the Quantitative Prediction of Sandstone Reservoir Quality, which was established in 2001 and has over 20 member companies. He also collaborates with the Fracture Research and Applications Consortium at the University of Texas at Austin and is a research fellow with the Bureau of Economic Geology. He has a PhD from the University of Illinois at Urbana-Champaign and a BA from Knox College.

The message is the medium 96

Christopher Liner is Chairman of the Department of Geosciences at the University of Arkansas and holds the Maurice F Storm Endowed Chair in Petroleum Geology. He served as 2014–15 SEG President, was the editor of *Geophysics*, and was the 2012 SEG Distinguished Instructor (*Elements of Seismic Dispersion*). He authors the long-standing occasional column *Seismos* in *The Leading Edge*, *Seismos* blog, and the *World Oil* magazine column *What's New in Exploration*. Liner's book projects include finalizing SEG's 3rd edition of *Elements of 3D Seismology* and co-authoring *The Art and Science of Seismic Interpretation* to be published by Springer. He teaches the EAGE short course *Carbonate Essentials: Pores to Prospects*. Liner's background is both academic and industrial. Eleven years of business experience includes Western Geophysical, Conoco, Golden Geophysical, and Saudi Aramco. He has held faculty positions at the universities of Tulsa, Houston, and, currently, Arkansas. His research interests are carbonate characterization and advanced seismic interpretation methods. His education includes the University of Arkansas (BSc in geology), University of Tulsa (MSc in geophysics), and a PhD in geophysics from the Colorado School of Mines working with the Center for Wave Phenomena.

Negative Q 64

John Logel is a geophysical consultant to various organizations as a mentor, teacher, and prospect reviewer. John's previous positions were as Chief Geoscientist North Sea for Talisman Energy Norge/UK in Aberdeen, Scotland, the lead geophysicist in Norway, and Senior Geoscience Advisor for North American Operations in Calgary, Alberta. Prior to Talisman, he has held several technical management and advising positions with Anadarko Canada and Petro-Canada in Calgary, and before that spent 19 years with Mobil on numerous assignments in Europe and North America. He holds a BSc and MSc from the University of Iowa.

Understand the shales to understand the reservoir 102

Ali Misaghi was educated in Iran and at the Norwegian University of Science and Technology. He has worked for Kayson Oil, Gas, and Energy Company (KOGC), Dana Energy, and the Research Institute of Petroleum Industry of Iran. He is currently a managing partner at Tavana Energy Company.

What is Gassmann's equation? 106

Dave Monk holds a PhD in physics from Nottingham University in the UK. He is currently director of geophysics at Apache Corporation. He is the author of over 100 technical papers or articles and a number of patents, has received Best Paper awards from the Society of Exploration Geophysicists (1992 and 2005) and the Canadian SEG (2002), and was recipient of the 1994 Hagedoorn Award from the European Association of Exploration Geophysicists (now EAGE). He served as President of the Society of Exploration Geophysicists (SEG) in 2012–13.

What is the speed of sound?   *108*

Rachel Newrick obtained undergraduate degrees at Victoria University of Wellington, New Zealand and her PhD in exploration seismology at the University of Calgary in 2004. Since 1992, her petroleum experience includes working with BHP Petroleum in Melbourne, Occidental Petroleum in Houston, Exxon Mobil in Houston, Veritas DGC and Nexen Inc in Calgary, Nexen Petroleum UK in London, and Cairn Energy in Edinburgh. Since leaving Cairn Energy in 2013, Rachel has continued to focus on frontier exploration as an independent consultant with Racian Ventures in Alberta, Canada. She is also a tutor for Nautilus predominantly teaching seismic interpretation at international locations. Rachel is the co-author of SEG Geophysical Monograph Series #13 *Fundamentals of Geophysical Interpretation* with Laurence Lines, contributed to *52 Things You Should Know About Geophysics*, and is involved with a number of geoscience technical societies. She was the CSEG President for 2015 and currently serves as a Past President.

Where did you get those logs?   *112*

Paola Vera de Newton received a petroleum engineering degree (2002) from the University of Zulia, Venezuela, and that same year joined Venezuelan state company PDVSA as a petroleum engineer. She received an industrial project management degree from the Rafael Belloso Chacín University, Venezuela. She is the technical advisor for petrophysics and rock physics at Rock Solid Images (RSI). She joined RSI in 2007, and since then she has mainly worked in the seismic petrophysics group. She has been involved in multiple conventional and unconventional studies worldwide. Some of her recent rock physics projects include West Africa, Williston Basin, Gulf of Mexico, Marcellus, Monterey, Eagle Ford, North Sea, and Barents Sea. She is a member of SEG, AAPG, EAGE, and SPWLA.

Ask your data first, not your model   *18*

Dick Plumb has 40 years experience in earth stress analysis and petroleum geomechanics in 25 countries. He spent 29 years developing the science, measurements, software, and methodology underpinning Schlumberger's geomechanics business. He retired from Schlumberger in 2010 and formed Plumb Geomechanics, LLC. He has a PhD in geophysics from Columbia University, New York.

The dynamic to static correction   *92*

Brian Russell started his career as an exploration geophysicist with Chevron in 1976, and worked for Chevron affiliates in both Calgary and Houston. He then worked for Teknica and Veritas in Calgary before co-founding Hampson-Russell in 1987 with Dan Hampson. Since 2002, Hampson-Russell has been a fully owned subsidiary of Veritas and Brian is currently vice president of Veritas Hampson-Russell. He is also an Adjunct Professor in the Department of Geology and Geophysics at the University of Calgary. Brian was President of the CSEG in 1991, received the CSEG Meritorious Service Award in 1995, the CSEG medal in 1999, and CSEG Honorary Membership in 2001. He served as chair of *The Leading Edge* editorial board in 1995, technical co-chair of the 1996 SEG annual meeting in Denver, and as President of SEG in 1998. In 1996, Brian and Dan Hampson were jointly awarded the SEG Enterprise Award, and in 2005 Brian received Life Membership from SEG. Brian holds a BSc in geophysics from the University of Saskatchewan, an MSc in geophysics from Durham University, UK, and a PhD in geophysics from the University of Calgary. He is a registered Professional Geophysicist in Alberta.

A primer on poroelasticity 14

Mark Sams is QI manager of Ikon Science Asia Pacific. He has more than 30 years' experience in academia and industry. He spent 12 years at Imperial College, London, where he received an MSc and PhD in geophysics and carried out postdoctoral research in rock physics. He moved to Malaysia in 1994 and joined Petronas Research working on AVO analysis and inversion. He then worked for Jason for 14 years and joined Ikon Science in early 2012. He specialises in rock physics and seismic reservoir characterization and has published and presented widely on both topics. Mark is an Associate Editor of *Geophysical Prospecting* and a member of EAGE and SEG.

The Hilterman approximation 94

Colin Sayers is a scientific advisor in the Schlumberger Seismics for Unconventionals Group in Houston. He entered the oil industry to join Shell's Exploration and Production Laboratory in the Netherlands in 1986, and moved to Schlumberger in 1991. His technical interests include rock physics, exploration seismology, reservoir geomechanics, seismic reservoir characterization, unconventional and fractured reservoirs, seismic anisotropy, borehole/seismic integration, stress-dependent acoustics, and advanced sonic logging. He is a member of the AGU, EAGE, GSH, SEG, and SPE, and a member of the research committee of the SEG. He has a BA in physics from the University of Lancaster, UK, a DIC in mathematical physics, and a PhD in physics from Imperial College, UK. In 2010 he presented the SEG/EAGE Distinguished Instructor Short Course *Geophysics under stress: Geomechanical applications of seismic and borehole acoustic waves*, and was chair of the editorial board of *The Leading Edge*. In 2013 he was awarded Honorary Membership of the Geophysical Society of Houston "in recognition and appreciation of distinguished contributions to the geophysical profession".

Texture matters 88

Marc Sbar got his PhD in earthquake seismology from Columbia University in 1972, and enjoyed a research and teaching career at Lamont-Doherty Earth Observatory and the University of Arizona until 1983. He then changed gears and spent 18 years as a geophysical specialist at BP, before moving to Phillips Petroleum in 2000, staying on at ConocoPhillips until 2009. Returning to academia, Marc recently moved to Tuscon, Arizona, impressing the next generation of University of Arizona geoscientists with the wonders of geophysics.

Why do we need rock physics?

Rob Simm is a seismic interpreter with a special interest in rock physics, AVO, and seismic inversion technologies. Rob's early career (1985–1999) was spent with British independent oil and gas exploration companies Britoil, Tricentrol, and Enterprise Oil, working in both exploration and production. An interest in applying rock physics in prospect generation and field development led him to set up his own consultancy, Rock Physics Associates Ltd, in 1999. His training course *The Essentials of Rock Physics for Seismic Amplitude Interpretation* is recognised worldwide. Rob is author (with Mike Bacon) of *Seismic Amplitude: An Interpreter's Handbook* (Cambridge University Press, 2014). Rob is currently principal geophysics advisor with Origo Exploration, a small independent oil and gas exploration company based in Stavanger, Norway.

Some do's and dont's

Christopher Skelt has worked in research and development, consulting, subsurface management and operations for several exploration and production companies and service providers. His current work with Chevron Energy Technology Company is about aligning geophysicists' and petrophysicists' perception of the subsurface.

Tad Smith is director of E&P technology geoscience for Apache in Houston. He has a PhD in geology from Texas A&M University and has worked at Amoco, BP, Veritas, and ConocoPhillips. He was the 2011 SEG North American Honorary Lecturer. From 2011–15 he was on the editorial board for *The Leading Edge*. He also served as President of the Geophysical Society of Houston during the 2014-15 term.

Some thoughts on the basics

Kyle T Spikes is an Associate Professor in the Department of Geological Sciences at the University of Texas, Austin. His research focuses on the development of techniques to integrate geological information with quantitative geophysical tools for seismic reservoir characterization. This area of research includes both forward and inverse problems, which combine rock physics, stochastic geological modelling, seismic inversion, and wavefield simulation.

Consider the error in the measurements

Sven Treitel grew up in Argentina and was educated at MIT where he graduated with a PhD in geophysics in 1958, before enjoying a long career at Amoco. Sven has published over 40 papers and is the recipient of numerous learned society awards, including the 1969 SEG Reginald Fessenden award, and in 1983 was awarded Honorary Membership of SEG. While his interests have been broad and varied, his main contribution to the field of geophysics has been to bridge the gap between signal processing theory and its application in exploration geophysics. He is the co-author of the definitive volumes *Geophysical Signal Analysis* (Prentice-Hall, 1980 & SEG, 2000) and *Digital Imaging and Deconvolution* (SEG, 2008). Although officially retired, Sven still lectures and consults widely.

Futoshi Tsuneyama is a deputy general manager of Idemitsu Kosan in Tokyo, Japan. He was manager of the Technical Evaluation Department of Idemitsu's E&P subsidiary until 2013. He received his MSc in geology and mineralogy from Niigata University in 1989. After working in the upstream oil industry for 14 years, his interest turned to rock physics so he decided to go back school to expand his technical background. Eventually, he received a PhD in geophysics from Stanford University in 2005, the year Steve Jobs made a speech at the commencement.

K M (Chris) Wojcik has an MSc in Sedimentology from Warsaw University and a PhD in sedimentary petrology and diagenesis from the University of Kansas. He joined Shell in 1991 as an exploration geologist in New Orleans and moved to the quantitative interpretation (QI) area in 1998 with Shell Angola and Shell Brazil. Later, he focused on the deployment of QI technologies in many deepwater basins and worked as the exploration QI specialist with Shell Nigeria in Lagos. Currently, Chris holds the position of Geophysical/QI Advisor with Shell E&P in Houston. Derisking low-saturation gas has been a keen interest throughout his career.

Fuyong Yan holds a PhD in exploration geophysics, with master degrees in petroleum geology and petroleum engineering. He had worked as a VSP processing geophysicist with CGG for four years. His current research is primarily involved with the Fluids/DHI Consortium. His research interests include poroelasticity of reservoir rocks, seismic anisotropy, and application of rock physics on seismic reservoir characterization.

Fernando Enrique Ziegler received a BSc in physics from the University of Texas at Austin and an MSc in geophysics from the University of Houston. He started his career in testing seismic applications for Petroleum Geo-Services and then moved to geophysical migration and processing quality control for GX Technology. Fernando eventually began to specialize in pore pressure prediction for Marathon Oil and Repsol. Currently, he is an independent geophysics consultant working on pore pressure, fracture pressure, wellbore stability, seal capacity, hydrocarbon column height estimates, rock physics, and geomechanics. He is a committee member (2007 to present) and Chair (2014 to present) of the AAPG/SEG Student Expo Committee, committee member of the Operators' Pore Pressure Forum (2015 to present), and member of the Committee on University and Student Programs at SEG (2008 to present). He is a member of SEG, AAPG, and SPE.

Pore pressure and everything else   *70*

Index

Published 2014

Available from *ageo.co/52geology*
& online bookstores worldwide.

$19 · £12 · €15

ISBN 978-0-9879594-2-3

"This new book is full of practical tips, commentary, and
advice from real geologists who have been there
and know what the science is all about."

Andrew D Miall — Professor of Geology, University of Toronto

"If you're right at the start of your career this volume will put
you in touch with the human side of a geoscientist's life in oil
and gas, and will entertain as well as enlighten you.
For anyone else, you may nevertheless enjoy the insights
into the preoccupations, predilections, and prejudices
of the life of an exploration geologist."

David E Smith — Professor of Geography, University of Oxford

Published 2015

Available from *ageo.co/52palaeo*
& online bookstores worldwide.

$19 • £12 • €15

ISBN 978-0-9879594-4-7

"This is sheer delight for the reader, with a great range
of short but fascinating articles; serious science
but often funny. Altogether brilliant!"

Professor Euan Clarkson, FRSE

University of Edinburgh

43354508R00077

Made in the USA
Middletown, DE
07 May 2017